The Skeleton Key of Mathematics

A Simple Account of Complex Algebraic Theories

D. E. Littlewood

DOVER PUBLICATIONS, INC.
Mineola, New York

Published in the United Kingdom by David & Charles, Brunel House, Forde Close, Newton Abbot, Devon TQ12 4PU.

Bibliographical Note

This Dover edition, first published in 2002, is an unabridged republication of the work first published by Hutchinson & Company, Limited, London, in 1949 and reprinted by Harper & Brothers, New York, in 1960.

Library of Congress Cataloging-in-Publication Data

Littlewood, Dudley Ernest.
The skeleton key of mathematics : a simple account of complex algebraic theories / D.E. Littlewood.
p. cm.
Originally published: London : Hutchinson & Co., 1949. (Hutchinson's university library. Mathematical and physical sciences ; no. 18)
Includes index.
ISBN 0-486-42543-6 (pbk.)
1. Algebra. I. Title.

QA155 .L5 2002
512—dc21

2002073492

Manufactured in the United States of America
Dover Publications, Inc., 31 East 2nd Street, Mineola, N.Y. 11501

CONTENTS

	Preface	*Page* 7
Chapter I	The Method of Abstraction	9
II	Numbers	15
III	Euclid's Algorithm	23
IV	Congruences	30
V	Polynomials	39
VI	Complex Numbers and Algebraic Fields	44
VII	Algebraic Integers, Ideals and *p*-adic Numbers	50
VIII	Groups	58
IX	The Galois Theory of Equations	65
X	Algebraic Geometry	77
XI	Matrices and Determinants	86
XII	Invariants and Tensors	94
XIII	Algebras	101
XIV	Group Algebras	107
XV	The Symmetric Group	115
XVI	Continuous Groups	123
XVII	Application to Invariants	131
	Index	137

PREFACE

THERE is a story in the Old Testament concerning the Tower of Babel, how men sought to build a tower that would reach the heavens, but were cursed with a confusion of tongues.

The story is not without relevance to the science of today, which aspires in some respects beyond the heavens. The curse of the confusion of tongues is no less apt. What scientist can read with interest a technical paper in a different branch of science from his own? What mathematician can read with profit research papers on a topic on which he has not specialized knowledge?

A specialist has been described as a man who knows more and more about less and less. But such specialized knowledge is of little real value unless it is co-ordinated with the body of general knowledge. But it is just this co-ordination which becomes increasingly difficult in modern times.

One great aid to such co-ordination is the existence of abstract principles which are common to the various branches of science and mathematics. These can have the effect of weaving together the separate specialized techniques. In this, the abstraction of algebra might play no trivial role.

Urgently required, however, are more books of an intermediate character. Books that go deeper than the popular science series, and give a real description of the specialized work that has been accomplished, but being intended for the general intelligent reader rather than the specialist, describe only the general contours and omit the troublesome details. Above all, they must *not* begin by assuming in the reader a knowledge of all the technical terms, words and devices, which in the more specialized books and papers form an effective, almost impenetrable, barrier to the unskilled reader.

With these views in mind this book has been written. It

is hoped that there is sufficient general descriptive account of the theories to catch and keep the interest of the general intelligent reader. It is also hoped that even the specialized mathematician will find something new in its pages.

D. E. LITTLEWOOD.

June, 1947.

CHAPTER I

THE METHOD OF ABSTRACTION

MOST people possessing a lock possess also a key that will open it. But to possess a key that will open other locks, strange locks never before seen, has a thrill of its own, and the idea of a skeleton key has lent zest to many a boy's reading of light detective fiction. And though modern locks such as the Yale, and locks with tumblers, are not susceptible to the use of such skeleton keys, yet the concept has never quite lost its glamour.

A locksmith may make fifty locks, each with its own key that will open none but the corresponding lock, but he can make one master key that will open every lock. This is because the opening mechanism is the same for each lock. One small portion of the key operates this mechanism, and this small portion must be present in every key. In the master key there is that small portion alone, joined by a thin bar to the shaft of the key. There is nothing in the master key that is not essential to every key, and because there is nothing redundant, there is nothing to get in the way to prevent the key from turning, in whichever of the fifty locks it is used.

The principle used in making a skeleton key is thus the concentration on the minimum effective part with the exclusion of everything redundant or non-essential. This austerity brings as a reward a vastly increased power and breadth of application.

The same principle in mathematics is called *abstraction* and it is perhaps the guiding motif of mathematics. Mathematics was indeed born of abstraction, for consider the very concept of number.

A trained sheepdog may perceive the significance of two, three or five sheep, and may know that two sheep and three sheep make five sheep. But very likely the knowledge would tell him nothing concerning, say, horses. A child learns that two fingers and three fingers make five fingers, that two beads and three beads make five beads. Then the irrelevance of the fingers or the beads, or the exact nature of the things that are counted becomes evident, and by the process of abstraction the universal truth that $2 + 3 = 5$ becomes evident.

At a much later stage it may be noticed that if unity is subtracted from the square of 5, to give $25 - 1 = 24$, then this number has two factors which are respectively $5 + 1 = 6$ and $5 - 1 = 4$. Similarly $6^2 - 1 = (6 + 1)(6 - 1)$ and $7^2 - 1 = (7 + 1)(7 - 1)$. It is clear that the proposition is true whatever number is used in the place of 5, 6 or 7. If then a new symbol, say x, is introduced which is not restricted to one number, but can represent any number, then the universal proposition can be written

$$x^2 - 1 = (x + 1)(x - 1).$$

This is a master key which enables one to express as a product of factors all the numbers, 8, 15, 24, 35, 48, 63, etc. The introduction of such a symbol x is the beginning of algebra.

With the advance of scientific knowledge, as the more obvious gaps are filled, enquiry tends to become more and more specialized. This specialization tends to restrict the scope of application of any discoveries. In mathematics this tendency may be offset by the abstraction which has the opposite broadening effect. The net result of these two influences is sometimes remarkable. Because of the specialization the immediate range and scope of a result may be severely restricted. But because of the depth of abstraction unexpected applications appear later which are extremely remote from the context, in a different branch of mathematics, or in a quite different subject.

This effect may be very beneficial, in that it has a tendency to integrate science which specialization tends to separate and divide.

A spectacular example of this influence at a distance is furnished by Einstein's General Theory of Relativity. Previously Riemannian Geometry had been a very specialized study whose main interest was its more general character, in that it included the usual Euclidean Geometry and even hyperbolic and elliptic Geometries as special cases. Its study was more difficult, but a technique had been developed. It was hardly anticipated that there would ever be an application that was not of a technical and specialized character. Then the subject suddenly provided a clue as to the nature of the universe, and in particular as to the nature of gravitational action at a distance. According to Einstein, the Geometry of the world was not Euclidean, but Riemannian, the law of gravitational force should be replaced by a geometrical law, the vanishing in free space of the first contraction of the curvature tensor, and the paths of the heavenly bodies were geodesics in the curved space time. The calculation was made possible by the previously developed technique of the geometers.

The instance of Einstein's General Relativity theory had its parallel some two and a half centuries earlier. At the time of Newton the theory of *conic sections* had been studied for 1600 years as pure mathematics, simply for the love of abstract knowledge. But because the paths of heavenly bodies were to the first approximation conic sections, this fact gave to Newton the clue which led to the inverse square law of gravitation.

The existence of such sensational applications is not necessarily the criterion of value of a mathematical result. Valuation is one of the most essential functions of living, but it is also the most controversial of topics. Some people would demand, as proof of the value of mathematics, the most practical and utilitarian of applications to our daily lives. Others react so violently to this demand for practical results that a famous Cambridge Don is said to have given the toast "To Pure Mathematics, and may it never be of any use to anyone." The demand for uselessness is not to be taken seriously, but serves to emphasize that real value exists in true knowledge which puts to scorn the test of practical applications.

In this context it is not without significance that in those Universities which give degrees in both Arts and Science, Mathematics is recognized as a subject either for an Arts or for a Science degree.

As an Art, Mathematics has its own standard of beauty and elegance which can vie with the more decorative arts. In this it is diametrically opposed to a Baroque art which relies on a wealth of ornamental additions. Bereft of superfluous addenda, Mathematics may appear, on first acquaintance, austere and severe. But longer contemplation reveals the classic attributes of simplicity relative to its significance, and depth of meaning.

What qualities would be shown in a mathematical proof that is elegant? The strictest economy in the initial assumptions and in the procedure of the proof would be a criterion, directness, and absence of all irrelevance. But these are the essential attributes of abstraction.

What qualities, again, would one expect to find in a really great mathematical discovery? One would certainly expect deep significance and great generality, "a light shining from the east which giveth light even unto the west". These again are the attributes of abstraction.

Returning to the standard of values of the man who insists on practical utility, he may fail to be impressed even by so spectacular an application as to Einstein's General Relativity theory. And indeed, this may be regarded in some measure as disappointing. Failing to explain electrical forces at the same time as gravitational forces, it seemed to tell but half the story, and the general insight which it promised to give into the nature of the universe was not altogether fulfilled. Perhaps the fruits of this theory are yet to be gathered.

But consider Einstein's much less ambitious Special Theory of Relativity. This sought to explain the fact that however and wherever it is measured, the velocity of light in vacuo always appears to be exactly the same. If a train moves at 60 miles an hour, and a car goes in the same direction at 20 miles an hour, then the train would appear to be moving relative to the car at only 40 miles an hour. Now the surface of the earth must be moving with considerable speed because of the rotation of the earth about its axis, and because of its motion round the sun. This should alter the apparent velocity of light to a degree that could easily be detected with the sensitive apparatus of the Michelson-Morley interferometer. But no such change could ever be detected.

The law of relative velocities must be wrong. To find the correct law it was necessary to solve the algebraic problem of finding the group of transformations which leave a velocity, i.e. the velocity of light, invariant. Fitzgerald and Lorentz found such a transformation. Einstein showed how the laws of mechanics must be modified to fit into the scheme.

The conclusions so far seem very abstract and impracticable, but one consequence of Einstein's theory made it necessary that energy should have both mass and weight. Thus if the spring of a piece of clockwork were wound up, the clockwork would become heavier, by an amount which could be exactly calculated though it would be too small for measurement.

The energy that is bound inside the nucleus of an atom, however, is vast and makes a measurable contribution to the weight of the atom. This energy can be estimated precisely from a knowledge of the atomic weights of the elements concerned. Thus since the atomic weight of hydrogen is 1·0081, and that of helium is 4·0039, it is clear that if four atoms of hydrogen are made to combine to form an atom of helium, then 1 part in 140 of the total mass is lost. This fraction is converted into energy. One ninth part of the weight of ordinary water consists of hydrogen. Hence if a one-ton projectile contains one ounce of water, the hydrogen from which is converted into helium, then 1 part in $16 \times 2240 \times 9 \times 140$, or about 1 part in 45 million of the mass of the projectile is converted into energy. Denote the mass of the projectile by M and the decrease in mass by m, so that $M = 45{,}000{,}000\ m$. The energy liberated is mc^2 where c is the velocity of light. If this energy is used to give the projectile a velocity V, the kinetic energy of the projectile is $\frac{1}{2} MV^2$, and thus

$$\tfrac{1}{2} MV^2 = mc^2$$

or

$$\begin{aligned} c^2 &= 22{,}500{,}000\ V^2, \\ c &= 4{,}500\ V. \end{aligned}$$

Since c is 200,000 miles per second, approximately, then V is 44 miles a second.

The energy released when the hydrogen in one ounce of water is converted into helium is sufficient then to give a projectile weighing one ton a velocity of 44 miles a second. This is sufficient to raise it to a height of 1000 miles above the earth's surface. Four ounces of water would be sufficient to remove the projectile completely from the earth's gravitational field.

No method of thus converting hydrogen into helium is yet known except on a microscopic scale, but as a possible future source of power the potentialities are vastly in excess of anything possible from Uranium fission. The Uranium fission and the atomic bomb project would hardly have been possible without the theoretical work on the nucleus and the detailed energy calculations in which this energy-mass relation derived from Einstein's theory is in constant application.

Such practical repercussions of what were purely theoretical, and in the first instance abstract algebraical investigations, must appeal to the most utilitarian and practical-minded of men.

Of that most abstract branch of knowledge, Mathematics, the most abstract subject is Algebra. Hence in Algebra especially one may hope to find, on the one hand, the richest examples of that austere beauty peculiar to Mathematics, and on the other hand the most potent theories and techniques possessing applications far removed from their context.

The earlier chapters of this book deal with topics whose main interests lie in their elegance and generality. Applications to the mundane affairs of daily life are not impossible, but are rare and improbable.

In the later chapters a different series of topics is discussed. While these have their own elegance, yet attention should rather be drawn to the depth of their significance which is such as to render possible, and almost probable, far-reaching consequences which may touch remote branches of knowledge and ultimately even affect our daily lives.

CHAPTER II

NUMBERS

A NECESSARY preliminary for any proper understanding of mathematics is to have a clear conception of what is meant by a number. When dealing with numbers most people refer to their own past experience in handling numbers, and this is, usually, not inconsiderable. Familiarity gives confidence in the handling, but not always an insight into the significance. The technique of manipulating numbers is learnt by boys and girls at a very tender age when manipulative skill is fairly easily obtained, but when the understanding is very immature. At a later stage, when the faculty of understanding develops, the technique is already fully acquired, so that it is not necessary to give any thought to numbers. Thus the immaturity of concept persists. To appreciate the significance of numbers it is necessary to go back and reconsider the ground which was covered in childhood.

Apart from specialized mathematicians, few people realize that, for example, the number 2 can have some half a dozen distinct meanings. These differences of meaning are reflected in the logical definitions of numbers.

Consider on the one hand two eggs, and on the other hand two pounds of butter. That these "two's" really mean something different becomes apparent if the fraction "one half" is substituted for "two". It is just as easy to take half-a-pound of butter as it is to take two pounds, but to remove a half of an egg from a basket containing two would present a problem. Breaking the egg and dividing the contents is no solution, for a broken egg is not the same as an egg. Again, it can be determined precisely by a process of counting that there are exactly two eggs in a basket. But the two concerning the pounds of butter has nothing to do with counting, unless by an unnecessary complication the butter is already weighed up in pounds, and the two in any case is only an approximation.

Again, consider one man who has £2 in a purse, and

another who has a credit of £2 in a banking account. The latter could possibly write a cheque and withdraw £3, leaving a negative balance of £1. But it is simply nonsense to consider the withdrawal of £3 from the purse so as to leave minus £1 therein.

In mathematical logic the cardinal integers are the first numbers to be defined. These are applicable to such problems as involve counting. The convenience which comes from the use of negative numbers makes it desirable to define a new system of numbers to which the + or − sign can be attached. It is these numbers which are involved in the example of the bank balance. An entirely new system of numbers has then to be defined if a meaning is to be attached to fractions, and these may be defined either with or without signs. Even this system of numbers seems incomplete, as a need arises for new numbers such as $\sqrt{a}$ or π which do not fit into the rational scheme. Hence is defined yet a fresh system of numbers, the *real numbers*, to include all these. Mathematicians go one step further to define *complex* numbers. These will be discussed in a later chapter.

A whole number such as 2 appears in each system, and though the rules of arithmetic are the same for each system, yet the exact meaning of 2 will vary according to the system of numbers which is being considered.

The method of the logical definitions of these numbers will now be described.

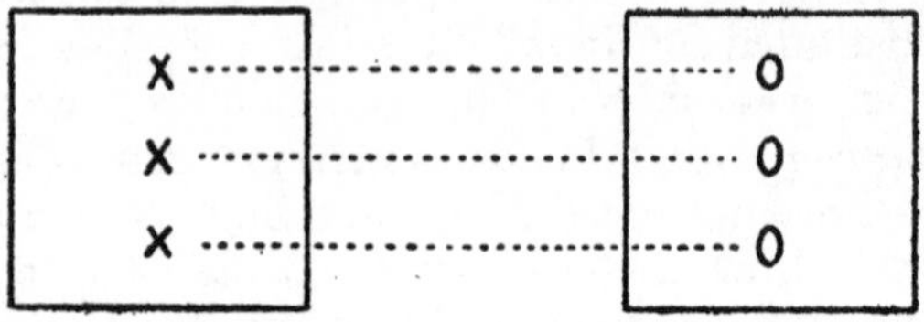

In the above squares there are respectively crosses and circles. Such sets are called *classes*. A class is any aggregate of entities which may be symbols as in this instance, objects such as apples or books, or abstract entities such as classes themselves. The entities are called *members* of the respective classes.

The class of crosses and the class of circles have this mutual property, that there exists a one-one correspondence between the crosses and the circles, each circle corresponding to exactly one cross, and vice versa. The classes are therefore said to be *similar*. The property which is possessed by all classes similar to either of these classes is what is usually associated with the number 3. A convenient method of abstracting a property common to a number of things is to consider the class of all the things which have this property.

A convenient definition of the cardinal number 3 is thus the class of all classes similar to the class of crosses. The class of crosses is said to possess this cardinal number. The cardinal number of the class of circles is obviously the same.

Addition of cardinal numbers is defined by forming the *sum class* of two classes, that is, the class whose members are members of either class. Provided the original classes have no common member the sum of the cardinal numbers is defined as the cardinal number of the sum class.

Multiplication is defined by considering the class of ordered pairs, one member from the first class and one from the second. The cardinal number of the class of ordered pairs is defined as the *product* of the cardinal numbers.

In this way a sound logical foundation can be constructed for the arithmetic of integers. It is as well to mention at this point the five fundamental laws of arithmetic:

Commutative addition law: $a + b = b + a$

Associative addition law: $a + (b + c) = (a + b) + c$

Commutative multiplication law: $a\ b = b\ a$

Associative multiplication law: $a\ (b\ c) = (a\ b)\ c$

Distributive law: $a\ (b + c) = a\ b + a\ c$

It is on these laws that the ordinary rules of arithmetic and algebra are based. After the integers are defined it is necessary to prove that the laws are obeyed. With every fresh type of number they must be proved all over again. The proofs in each case are quite simple and straightforward, and will be passed over here without further comment.

The integers so far have no signs, and the impossibility of subtracting say 3 from 2 is sometimes inconvenient. This inconvenience is felt especially by a man with £2 in a purse who needs very urgently £3. He could evade the difficulty perhaps by keeping his money in a banking account. Then having a balance of £2 he could write a cheque for £3 and leave a negative balance of £1.

Consider how this is possible in a banking account but not with money in a purse. The bank associates *two* totals with each customer's account, the total of moneys credited and the total of moneys withdrawn. The net balance is then regarded as the same if, for example, the credit amounts to £102 and the debit £100, as if the credit were £52 and the debit £50. If the debit exceeds the credit the balance is negative.

This model is adopted in the definition of the *signed integers*. Consider pairs of cardinal numbers (a, b) in which the first number corresponds to the debit, and the second to the credit. A definition of equality is adopted such that

$$(a, b) = (c, d)$$

if and only if $a + d = b + c$.

Addition is defined by the rule

$$(a, b) + (c, d) = (a + c, b + d),$$

and multiplication by

$$(a, b) \times (c, d) = (a\,d + b\,c,\ a\,c + b\,d),$$

then the pair (o, a) or any equivalent pair is denoted by a, and the pair (a, o) or any equivalent pair by $-a$.

Alternatively, the signed integers could be regarded as a transition from one cardinal integer to another, e.g. the transition $100 \rightarrow 102$, or $50 \rightarrow 52$ is denoted by $+2$, and the transition $102 \rightarrow 100$ by -2. The definition as a pair of cardinal integers is well adapted to this interpretation.

If addition is replaced by multiplication the same procedure leads to another generalization of numbers, this time to *rationals*. This generalization can be made either before or after the generalization to signed integers.

If a, n are integers (cardinal or signed), then the equation $a\,x = n$ may or may not have a solution. If there is a solution it is unique and one writes

$$x = \frac{1}{a} \times n \text{ or } x = \frac{1}{a} \text{ of } n,$$

then

$$b\,x = \frac{b}{a} \times n \text{ or } \frac{b}{a} \text{ of } n.$$

Considering the arithmetic of these operators $\frac{b}{a}$, clearly

$$\frac{b}{a} \text{ of } n + \frac{d}{c} \text{ of } n = \frac{b\,c + a\,d}{a\,c} \text{ of } n,$$

and

$$\frac{b}{a} \times \frac{d}{c} \text{ of } n = \frac{b\,d}{a\,c} \text{ of } n.$$

The operators can be added and multiplied and the resulting operators are independent of the number n operated on, provided only that it is such a number that the equation

$$a\,c\,x = n$$

has a solution.

It is therefore possible to ignore the number n and consider only the operators themselves, which are called *rational numbers* or *rationals*.

Logically the rational is defined as an ordered pair of numbers (cardinal or signed), (b, a), which is written for convenience $\frac{b}{a}$ or b/a. The definition of equality is taken as

$$\frac{b}{a} = \frac{d}{c}$$

if and only if $b\,c = a\,d$.

Addition is defined by the equation

$$b/a + d/c = (b\,c + a\,d)/a\,c,$$

and multiplication by

$$b/a \times d/c = b\,d/a\,c.$$

The rationals fill in the gaps between the integers to some extent, and seem adequate for most practical purposes, but are still not general enough for every occasion.

Thus if the side of a square is one foot, it is known that the area of the square on a diagonal as side is two square feet. This gives for the length of the diagonal a distance x such that

$$x^2 = 2.$$

Unfortunately no such rational number exists, for if

$$\frac{p^2}{q^2} = 2$$

where p and q have no common factor, the equation $p^2 = 2q^2$ implies that p is an even number so that p^2 is divisible by 4 and hence q^2 is divisible by 2. Thus q must be an even number and p and q must have a common factor 2, which is a contradiction.

No rational number, then, can satisfy $x^2 = 2$. There are still gaps in the number system of rationals, and to eliminate these gaps *real numbers* are defined.

Although no rational can be found to satisfy $x^2 = 2$, yet it is possible to separate the rationals into two classes, those which are too small and those which are too large. Thus 7/5 is too small since 49/25 is less than 2, but 17/12 is too large since 289/144 is greater than 2. Thus if a new number $\sqrt{2}$ is defined to satisfy $x^2 = 2$, an exact position amongst the ordered rationals can be assigned to it.

Similarly, if by any rule the set of rationals can be separated into two classes, an L-class and an R-class, such that every rational of the L-class is less than every rational of the R-class, then either there is a rational which separates the L-class from the R-class, or else a new number can be defined to mark this partition.

Mathematically the whole system of such numbers is defined in one operation by defining a *real number* to be a partition of the rationals into an L-class and an R-class with every rational of the L-class less than every rational of the R-class. Such a system includes all the roots of positive rationals and all such transcendental numbers as π and e.

No further generalization of number is possible which maintains the property of order. One further generalization is usual in mathematics, however, namely to complex numbers, which will be mentioned in a later chapter. The system is then, for almost all purposes, complete.

It should be noticed that the definition of real numbers depends entirely on the concept of *order*, of one rational being less than another. If a different system of ordering numbers is employed, the corresponding generalization is to an entirely different system of numbers with quite different properties. The p-adic numbers described in Chapter VII exemplify this.

Different theories and different branches of mathematics deal with different systems of numbers. Analysis deals almost entirely with the real numbers and their further generalization, the complex numbers. But in algebra there are theories concerning the integers and theories concerning the rationals. The theory of algebraic numbers concerns certain extensions intermediate between the rationals and the real numbers.

There is one very curious aspect of these number systems which is worth remarking. From a practical aspect the extension from integers to rationals is far more important than the extension from rationals to reals. The introduction between two integers p and $p + 1$ of a whole series of intermediate rationals appears a very considerable and useful extension, but the irrationals seem hardly necessary from a practical point of view. It is always possible to get a rational approximation that is "near enough".

Thus for the irrational number π representing the ratio of the circumference of a circle to the diameter, the rational 22/7 is often sufficiently accurate. If it is not, a nearer rational approximation is 3·14159. For more accuracy a rational approximation up to more than 500 decimal places has been calculated. The error here would be less in proportion than a single odd atom in a whole universe. So the concept of the real numbers to include these irrationals seems almost superfluous.

Nevertheless the extension from rationals to reals is actually a much vaster extension than the extension from integers to rationals. For it can be shown that the number of rationals is the same as the number of integers, but the number of reals is definitely greater.

The complete set of rationals p/q can be written down in a sequence by writing down first those rationals for which $p + q = 2$, then $p + q = 3, 4, 5$, and so on, thus

$$\frac{1}{1}, \frac{1}{2}, \frac{2}{1}, \frac{1}{3}, \frac{2}{2}, \frac{3}{1}, \frac{1}{4}, \frac{2}{3}, \frac{3}{2}, \frac{4}{1}, \frac{1}{5}, \frac{2}{4}, \ldots$$

From this sequence those rationals which are not in their lowest terms, e.g. 2/2, 2/4, 3/3, 4/2, . . . can be deleted. By numbering these rationals in order, 1, 2, 3, 4, etc., a one-one correspondence is obtained between all the rationals and all the integers. Hence the number of rationals is the same as the number of integers.

It is impossible to do the same with the real numbers. It is sufficient to take the reals between 0 and 1.

If possible let the reals between 0 and 1 be arranged in a sequence

$$x_1, x_2, x_3, x_4, \ldots,$$

and let the expressions for these as infinite decimals be

$$x_1 = \cdot a_1\ b_1\ c_1\ d_1 \ldots$$
$$x_2 = \cdot a_2\ b_2\ c_2\ d_2$$
$$x_3 = \cdot a_3\ b_3\ c_3\ d_3$$

$$\cdot \qquad \cdot \qquad \cdot \qquad \cdot \qquad \cdot$$

Then to find a real number which is not included in this sequence it is only necessary to consider an infinite decimal $\cdot p_1 p_2 p_3 p_4 \ldots$ such that p_1 differs from a_1, p_2 differs from b_2, p_3 from c_3, p_4 from d_4 and so on.

All the real numbers between 0 and 1, therefore, cannot be arranged in a sequence. The set of integers and the set of rationals are said to be *countable* since they can be arranged in a sequence, but the real numbers are *not countable*.

It is pertinent to enquire why it is necessary to introduce real numbers since these constitute so vast an extension of the rationals, apparently to so little effect, since to every real number, one can obtain a rational approximation to any degree of accuracy.

The necessity for the real numbers is illustrated by an important class of theorems called *existence theorems*. A query often arises, does there exist a number with such and such a property? With rationals the answer is often "no", whereas with real numbers the answer would be "yes". To make sure that a number will always be existent and ready when it is required, this vast extension from rationals to reals is necessary.

CHAPTER III

EUCLID'S ALGORITHM

MOST schoolboys learn a device for finding the highest common factor of two numbers. Thus to find the highest common factor (H.C.F.) of 119 and 623, divide the larger by the smaller to obtain a remainder less than 119.

$$623 - 5 \times 119 = 28$$

then any factor of 623 and 119 must also divide $623 - 5 \times 119$ or 28. The H.C.F. of 623 and 119 is also the H.C.F. of 119 and 28. Proceeding similarly

$$119 - 4 \times 28 = 7$$
$$28 - 4 \times 7 = 0$$

Thus 7 is the required H.C.F.

There appears nothing remarkable to discriminate this procedure from the many other artifices and devices taught in an elementary mathematical course. The following difference exists, however. The value of the above procedure is not exhausted by its use in finding the H.C.F. of two numbers. It has greater generality. Almost exactly the same procedure can be used to find the H.C.F. of two algebraic polynomials. But of far more importance is the depth of consequence which follows. It is the first part of an algorithm which Euclid obtains in his tenth book. The full algorithm may be said to be the foundation of several branches of mathematics.

The second part of the algorithm is best illustrated in the case when the numbers have no common factor, so that the H.C.F. is unity. In finding the H.C.F. of 17 and 49 the work would be

$$49 - 2 \times 17 = 15$$
$$17 - 15 = 2$$
$$15 - 7 \times 2 = 1$$

The last remainder is 1, which is the H.C.F. The last equation expresses 1 in terms of the two preceding remainders, 2 and 15. The preceding equation expresses 2 in terms of the preceding remainder 15 and the smaller of the original numbers, i.e. 17. The first equation expresses 15 in terms of the original numbers 17 and 49. Thus 1 can be expressed in terms of 17 and 49.

$$1 = 15 - 7 \times 2 = 15 - 7\,(17 - 15)$$
$$= 8 \times 15 - 7 \times 17 = 8\,(49 - 2 \times 17) - 7 \times 17$$
$$1 = 8 \times 49 - 23 \times 17.$$

The procedure can be followed for any pair of integers. The result can be stated thus:

If p *and* q *are integers there exists an integer* h *called the H.C.F. of* p *and* q *such that* h *divides* p *and* q, *and also integers* P, Q *can be found such that*

$$P\,p - Q\,q = h.$$

The special case when $h = 1$ is most important.

If p *and* q *have no common factor the integers* P, Q *can be found such that*

$$P\,p - Q\,q = 1.$$

The former result constitutes Euclid's Algorithm. The algorithm not only shows that P and Q exist, but also shows how they can be calculated.

The first application of this result is very theoretical and may seem of little practical importance. It concerns the uniqueness of factorization.

It may be possible to factorize a number, e.g. 72, in many ways, thus $72 = 8 \times 9 = 6 \times 12$. Suppose then that each factor is factorized and so on until each factor is undecomposable into factors, or *prime*. Thus

$$72 = 8 \times 9 = 4 \times 2 \times 3 \times 3 = 2 \times 2 \times 2 \times 3 \times 3$$

Proceed in a similar way with a second factorization of 72, e.g.

$$72 = 6 \times 12 = 2 \times 3 \times 2 \times 6 = 2 \times 3 \times 2 \times 2 \times 3$$

Now it is apparent that in factorizing 72 by either method we arrive finally at the same set of prime factors. But will this always happen? Could it not be that in factorizing a number by different ways entirely different sets of prime factors would result?

The answer is obtained from Euclid's Algorithm which can be used to show that for the ordinary integers, factorization into prime factors must always be unique. The proof follows from the following theorem:

If p is a prime and p divides $n = uv$, then either p divides u or p divides v.

If p does not divide u, then there are integers P, V such that

$$Pp - Vu = 1$$

and

$$Ppv - Vuv = v.$$

But since p divides Ppv and also $Vuv = Vn$, then it follows that p divides v.

The unique factorization theorem follows, for if p is one of the prime factors in the first prime factorization it must divide one of the factors in the second factorization, i.e. coincide with one of the factors. Each factor in the first must coincide with a factor in the second, and uniqueness follows.

The theorem is not so trivial as it at first appears. It is by no means obvious that prime factorization should in this way be unique. In fact we shall see in a later chapter that there exist domains of algebraic integers which resemble in many ways and have many of the properties of the ordinary cardinal integers. But in some of these domains Euclid's Algorithm is not applicable, since there is no way of ensuring that the remainder, on dividing one integer by another, should be less than the divisor. In these domains it is possible to factorize a number in different ways into prime factors. In order to restore unique factorization the concept of the "ideal" is introduced and the number is then expressed, not as a product of prime numbers, but as a product of prime ideals, when the uniqueness theorem again becomes valid.

Another use of Euclid's Algorithm is concerned with partial fractions. In elementary arithmetic it is common to express the sum of several fractions as a single fraction with a compound denominator, thus

$$\frac{1}{3} + \frac{2}{5} - \frac{1}{7} = \frac{35 + 42 - 15}{105} = \frac{62.}{105}$$

For some purposes it is easier to deal with the separate fractions, however, where the denominators are prime or are powers of a prime, than with the single fraction. But it is not nearly so easy to express, say, 43/105 as a sum of thirds and fifths and sevenths. The only method of accomplishing this apart from trial and error is by the use of Euclid's Algorithm.

Factorize the denominator into two factors

$$105 = 7 \times 15$$

Euclid's Algorithm with 7 and 15 gives

$$15 - 2 \times 7 = 1,$$

thence dividing by 105

$$\frac{1}{7} - \frac{2}{15} = \frac{1}{105}$$

Treat the denominator 15 in the same way

$$15 = 5 \times 3,$$

$$2 \times 3 - 5 = 1,$$

$$\frac{2}{5} - \frac{1}{3} = \frac{1}{15}$$

$$\frac{2}{15} = \frac{4}{5} - \frac{2}{3}$$

Thus

$$\frac{1}{105} = \frac{1}{7} - \frac{4}{5} + \frac{2}{3}$$

Thence

$$\frac{43}{105} = \frac{43}{7} - \frac{172}{5} + \frac{86}{3} = 6\frac{1}{7} - 34\frac{2}{5} + 28\frac{2}{3}$$

$$\frac{43}{105} = \frac{1}{7} - \frac{2}{5} + \frac{2}{3}$$

A similar problem arises with polynomials, when, for example, an expression such as $\frac{(X^2 + 1)}{X^3 - X}$ is required to be expressed in the form $\frac{A}{X} + \frac{B}{X - 1} + \frac{C}{X \quad 1}$ Euclid's Algorithm can be used, but in this case it is not the only or even the usual method. It is necessary to use it, however, to prove that an expression in partial fractions always exists.

Applications of the algorithm to congruences both in integers and in polynomials will be described in later chapters. For the present we will turn to a beautiful generalization of the procedure which is realized in the theory of continued fractions.

Generally speaking, computational methods can only be used with rational numbers. When dealing with irrationals it is usual to take a rational approximation. More often than not decimals are used, the number of decimal places taken being dependent on the degree of accuracy required. But the use of decimals does not always give the simplest rational approximation. Clearly the fraction $1/3$ is simpler than the corresponding decimal $0{\cdot}333333\ldots$ Therefore it is of interest to study the best rational approximations. There will be a series of best approximations according to the degree of accuracy required.

Suppose we start with a rational approximation p/q to a greater accuracy than is desired. It is required to find a simpler fraction p'/q' which will give as close an approximation as possible to p/q. The denominator of the difference $\dfrac{p}{q}-\dfrac{p'}{q'}$ can be no greater than qq'. Hence the difference cannot be less than $1/qq'$.

We therefore seek a simpler fraction p'/q' such that

$$\frac{p}{q}-\frac{p'}{q'}=\pm\frac{1}{qq'},$$

which gives

$$q'p-p'q=\pm 1.$$

The numbers p', q' are just those obtained in Euclid's Algorithm.

Consecutive simpler approximations can be found in the same way from p'/q', and from the subsequent fractions obtained. There is no need, however, to go through the algorithm more than once.

If the consecutive quotients in executing the algorithm

with p, q, are respectively $a_1, a_2, a_3, \ldots a_n$ then it may be seen that

$$\frac{p}{q} = a_1 + \cfrac{1}{a_2 + \cfrac{1}{a_3 + \ldots \quad + \cfrac{1}{a_n}}}$$

This expression is called a *continued fraction*. For economy of space it is usually written in the form

$$a_1 + \frac{1}{a_2 +} \frac{1}{a_3 +} \frac{1}{a_4 +} \ldots + \frac{1}{a_n}$$

If in this continued fraction the last term $\frac{1}{a_n}$ is omitted, it can be shown that the remainder of the fraction, when reduced to the form of an ordinary fraction, gives the first approximation p'/q' as obtained from Euclid's Algorithm. It follows that the second approximation p''/q'' is obtained by omitting the last two terms $\frac{1}{a_{n-1} +} \frac{1}{a_n}$, and so on. This continued fraction thus gives the whole series of rational approximations as obtained by the above described method using Euclid's Algorithm.

The fraction

$$a_1 + \frac{1}{a_2 +} \frac{1}{a_3 +} \ldots \frac{1}{a_r}$$

is called the rth convergent of the continued fraction. Suppose that when brought to ordinary quotient form it is $\frac{P_r}{Q_r}$. It can be shown that the numerators and denominators satisfy the recurrence relations

$$P_r = a_r P_{r-1} + P_{r-2},$$
$$Q_r = a_r Q_{r-1} + Q_{r-2}.$$

It has been stated that the difference between P_n/Q_n and P_{n-1}/Q_{n-1} is $1/Q_n Q_{n-1}$. Sylvester has written some very elegant papers concerning these continued fractions among which he shows that the difference between P_n/Q_n and P_{n-r}/Q_{n-r} is $N_r/Q_n\,Q_{n-r}$ where N_r is the numerator of the $(r-1)$th convergent of reversed continued fraction

$$a_n + \frac{1}{a_{n-1}+}\;\frac{1}{a_{n-2}+}\;\frac{1}{a_{n-3}+}\;\cdots\;\frac{1}{a_1}$$

The number π representing the ratio of the circumference of a circle to its diameter is of continuous computational importance. The continued fraction for π has been worked out to a large number of terms. The following are the first 12 terms:

$$= 3 + \frac{1}{7+}\;\frac{1}{15+}\;\frac{1}{1+}\;\frac{1}{292+}\;\frac{1}{1+}\;\frac{1}{1+}\;\frac{1}{1+}\;\frac{1}{2+}\;\frac{1}{1+}\;\frac{1}{3+}\;\frac{1}{1+}.$$

The following are the first five convergents:

3; 22/7; 333/106; 355/113; 103,993/33,102.

The value 22/7 is often employed in elementary mensuration. The value 333/106 is not nearly so accurate as the next convergent 355/113. This is very accurate, as the next denominator 292 is a large number. Since the convergents are consecutively greater than and less than the complete continued fraction, the error in using 335/113 for π is less than the difference between this and the next convergent, which is one in $113 \times 33{,}102$ or 1/3,740,526.

The continued fraction will converge more rapidly if in the consecutive divisions the *nearest* integer is taken instead of the greatest integer *less than* the remainder of the fraction. Some of the remainders thus become negative, but the denominator 1 never appears. This has the effect of cutting out the comparatively inaccurate convergents such as 333/106. Following this practice the 12 terms given above for π may be compressed into 7 in the form

$$3 + \frac{1}{7+}\;\frac{1}{16-}\;\frac{1}{294-}\;\frac{1}{3-}\;\frac{1}{4-}\;\frac{1}{5-}\;\cdots$$

CHAPTER IV

CONGRUENCES

WHEN the hour hand of a clock goes beyond the twelve, it starts at one again. To go forward four hours or to go forward sixteen hours are equivalent so far as the final position of the hands is concerned. For this purpose then four is equivalent to sixteen. We say that four *is congruent to* sixteen *to modulus* twelve.

This is written.

$$4 \equiv 16 \pmod{12}.$$

In general p is congruent to q (mod n) if and only if $(p - q)$ is divisible by n.

$p \equiv q \pmod{n}$ if and only if $(p - q) = m\, n$. Every integer p is congruent (mod n), to one of the numbers, 0, 1, 2, . . . , $(n - 1)$. The particular number is found by dividing p by n until the remainder is less than n and considering this remainder, which is called the *residue* (mod n).

The most interesting case is when the modulus is a prime number, and for the time being we will consider congruences to a prime modulus, which will be denoted by p.

We define the product and the sum of two residues (mod p) to be the residue to which the product or sum of the numbers is congruent. In this way we obtain a number system which has many of the properties of the ordinary integers. Thus addition and multiplication are defined and these obey the commutative, associative and distributive laws.

But in one respect the residues to a prime modulus resemble the rational numbers rather than the integers, for division is always possible, except by zero.

Thus to modulus seven,

since $4 \times 2 = 8 \equiv 1 \pmod{7}$

therefore $1 \div 2 \equiv 4 \pmod{7}$

and similarly
$$1 \div 3 \equiv 5,$$
$$1 \div 4 \equiv 2,$$
$$1 \div 5 \equiv 3,$$
$$1 \div 6 \equiv 6 \pmod{7}$$

Division by 7 is, of course, impossible, since $7 \equiv 0 \pmod 7$, and this is equivalent to division by zero.

For the general prime modulus p, division is accomplished by the use of Euclid's Algorithm, which suffices to prove that division except by zero is always possible.

It is required to find an expression for $a \div b \pmod p$, that is to say, to find a residue x such that

$$b\,x \equiv a \pmod p.$$

Going through the process of Euclid's Algorithm for b and p suppose that

$$B\,b - P\,p = 1.$$

Then $$B\,b \equiv 1 \pmod p,$$
and $$a\,B\,b \equiv a \pmod p.$$
Thus $$x \equiv a\,B \pmod p.$$

We have here a remarkable system which, like the system of cardinal integers, obeys the five laws, commutative, associative and distributive, but unlike the cardinal integers, has a finite number of elements only, and is also closed to division except by zero.

Just as with the real numbers, an equation of degree r cannot have more than r roots, as we shall now show.

Firstly it was shown by Euclid's Algorithm that if $x\,y$ is divisible by p then either x or y is divisible by p. Hence if $x\,y \equiv 0 \pmod p$, then either $x \equiv 0 \pmod p$ or $y \equiv 0 \pmod p$.

Next if $f(x)$ is a polynomial of degree r and $f(\alpha) \equiv 0 \pmod p$ then $f(x) \equiv (x - \alpha)\,g(x)$ where $g(x)$ is of degree $(r - 1)$. For clearly $f(x) = (x - \alpha)\,g(x) + R$ where R is independent of x. Putting $x = \alpha$, $R \equiv 0$ since $f(\alpha) \equiv 0$.

Thus if $f(x) \equiv 0$ has roots $\alpha_1, \alpha_2, \ldots, \alpha_n$, then $f(x) \equiv A\,(x - \alpha_1)(x - \alpha_2)\ldots(x - \alpha_n) \equiv 0$, where A is of degree 0, i.e. a number. If now $f(\beta) \equiv 0$, then $A(\beta - \alpha_1)(\beta - \alpha_2)\ldots(\beta - \alpha_n) \equiv 0$, and one of the factors must be congruent to zero. Since A cannot be congruent to zero, for this would make the whole polynomial $f(x)$ identically congruent to zero, it follows that β must be congruent to one of the roots α. Thus the congruence cannot have more than n roots.

Of the $p - 1$ non-zero residues mod p, namely, $1, 2, 3, \ldots, p - 1$, the first and last satisfy $X^2 \equiv 1$. The others can be grouped in pairs so that the product of each pair is congruent to unity. Thus to modulus 11,

$$2 \times 6 \equiv 1,\ 3 \times 4 \equiv 1,\ 5 \times 9 \equiv 1,\ 7 \times 8 \equiv 1.$$

It follows that the product of the set of $p - 1$ non-zero residues must be congruent to -1, to modulus p. This is *Wilson's Theorem. If* p *is prime, then* $(p - 1)! + 1 \equiv 0 \pmod{p}$.

The notation $n!$ is used for "factorial n", or the product of the first n cardinal integers.

If r is a non-zero residue (mod p), then $a\,r$ and $b\,r$ cannot be congruent unless $a \equiv b$.

Thus the set of $p - 1$ non-zero residues

$$r,\ 2r,\ 3r,\ \ldots,\ (p-1)\,r$$

must be all distinct and congruent in some order to the $(p - 1)$ non-zero residues

$$1,\ 2,\ 3,\ \ldots,\ (p-1).$$

Forming the product in each case,

$$r^{p-1}(p-1)! \equiv (p-1)! \pmod{p},$$

and thus

$$r^{p-1} \equiv 1 \pmod{p}.$$

Multiplying by r so that the congruence remains true when $r \equiv 0$, it follows that every residue satisfies

$$r^p \equiv r \pmod{p}.$$

This is a remarkable equation, for it is satisfied by every number of the system. There is obviously no analogy with ordinary real numbers.

Notice that whereas every number of the system can be reduced by a multiple of p without altering the residue, so every index of a power can be reduced by a multiple of $(p - 1)$. Hence every equation, or *congruence* as it is called, when the equals sign $=$ is replaced by the congruence sign $\equiv$, can be expressed as an equation of degree less than or equal to p. This makes the number of independent equations finite in number. It remains to be seen whether all these equations have roots, or whether there are equations with no roots just as the equation $x^2 + 1 = 0$ has no roots over the real numbers. Considering the case of quadratic equations we shall see that, setting aside the equations with double roots, that is, the equations which are congruent to an exact square, exactly half of the remaining equations have a pair of distinct roots, the other half have no roots.

The quadratic congruence

$$a\,x^2 + b\,x + c \equiv 0 \pmod{p},$$

can be solved by exactly the same method of completing the square, as can the ordinary quadratic equation. Thus

$$\begin{aligned} 4a^2 x^2 + 4ab\,x &\equiv -4\,ac \\ 4a^2 x^2 + 4ab\,x + b^2 &\equiv b^2 - 4\,ac \\ (2\,a\,x + b)^2 &\equiv b^2 - 4\,ac. \end{aligned}$$

Hence, provided that the square root can be found, the formula for the roots of a quadratic congruence is exactly the same as the formula for the roots of a quadratic equation

$$x \equiv -b \pm \sqrt{(b^2 - 4\,ac)}/2a.$$

The existence or non-existence of the roots depends upon whether the *discriminant* of the quadratic, namely $(b^2 - 4\,ac)$ is congruent to an exact square or not.

A residue is called a *quadratic residue* (mod p) if it is congruent to an exact square. Otherwise it is a *quadratic non-residue*. There are $(p - 1)$ non-zero residues (mod p). By squaring these $(p - 1)$ squares are obtained. But these squares are congruent in pairs, for $a^2 \equiv (p - a)^2 \pmod{p}$. Further, if $a^2 \equiv b^2$, then either $a \equiv b$ or $a + b \equiv 0 \pmod{p}$, so that no more than two distinct residues have congruent squares. It follows that there are, excluding zero, exactly $\frac{1}{2}(p - 1)$ quadratic residues (mod p), and the other $\frac{1}{2}(p - 1)$ residues are quadratic non-residues.

Since the only even prime is 2, it is assumed that p is odd. Suppose $p = 2\,\nu + 1$. Then it has been shown that every non-zero residue satisfies $x^{2\nu} \equiv 1 \pmod{p}$. If r is a quadratic residue, $r \equiv s^2$, and since $s^{2\nu} \equiv 1$, it follows that all quadratic residues satisfy $r^\nu \equiv 1$. But since the congruence $x^\nu \equiv 1$ has only ν distinct roots, these must be all the quadratic residues. Hence all quadratic non-residues must satisfy $x^\nu \equiv -1 \pmod{p}$.

It is convenient to let the symbol $\left(\frac{r}{p}\right)$ denote $+1$ if $r^\nu \equiv +1$ and -1 if $r^\nu \equiv -1$. Then clearly

$$\left(\frac{rs}{p}\right) = \left(\frac{r}{p}\right)\left(\frac{s}{p}\right)$$

To determine whether any number is a quadratic residue

(mod p), it is thus sufficient to know whether the prime numbers are quadratic residues.

This in turn can be deduced from a remarkable theorem called the *theorem of quadratic reciprocity*. This shows that if p and q are odd primes then the condition that p is a quadratic residue (mod q) is connected with the condition that q is a quadratic residue (mod p). In fact if either p or q is of the form $(4r + 1)$, then either p and q are both quadratic residues with the other as modulus, or both quadratic non-residues. If both p and q are of the form $(4r - 1)$, then one is a quadratic residue, the other a quadratic non-residue. Thus

$$\left(\frac{p}{q}\right)\left(\frac{q}{p}\right) = (-1)^{\frac{1}{4}(p-1)(q-1)}$$

The result is so simple that one looks for a simple proof. The known proofs, however, are rather intricate and involved and will not be attempted here.

Together with the auxiliary equations which determine whether 2 and -1 are quadratic residues, namely

$$\left(\frac{2}{p}\right) = (-1)^{(p^2-1)/8}, \quad \left(\frac{-1}{p}\right) = (-1)^{(p-1)/2},$$

the quadratic reciprocity theorem is sufficient to determine whether any residue is a quadratic residue or non-residue.

Thus to determine whether 119 is a quadratic residue (mod 257), the calculation is as follows:

$$\left(\frac{119}{257}\right) = \left(\frac{7}{257}\right)\left(\frac{17}{257}\right)$$

Since 257 is of the form $(4r + 1)$,

$$\left(\frac{7}{257}\right) = \left(\frac{257}{7}\right) = \left(\frac{5}{7}\right) = \left(\frac{7}{5}\right) = \left(\frac{2}{5}\right) = -1.$$

Also

$$\left(\frac{17}{57}\right) = \left(\frac{257}{17}\right) = \left(\frac{2}{17}\right) = (-1)^2 = +1.$$

Thus

$$\left(\frac{119}{257}\right) = -1 \times 1 = -1, \text{ and } 119 \text{ is not a quadratic residue.}$$

Prime Power Moduli

With a prime modulus, if $xy \equiv 0 \pmod{p}$, then either $x \equiv 0$ or $y \equiv 0 \pmod{p}$. This conclusion is no longer true if the modulus is not prime, for if $n = xy$, then $xy \equiv 0 \pmod{n}$ and $x \not\equiv 0$, $y \not\equiv 0$, $\pmod{n}$. The residues x, y are called *factors of zero* $\pmod{n}$. In particular, since $xy \equiv 0$ they are *associated* factors of zero.

If the modulus is p^n where p is prime, then a residue will be a factor of zero if and only if it is congruent to zero $\pmod{p}$. The residues which are not factors of zero are called non-singular residues. A factor of zero is a singular residue.

Division, in general, is not possible by a singular residue. If it is possible it is not unique. Thus

$$6\,x \equiv 4 \pmod{9}$$

has no solution, but

$$6\,x \equiv 3 \pmod{9}$$

has the three solutions 2, 5, 8.

For a non-singular residue, however, division is possible and is unique. The solution of the congruence

$$a\,x \equiv b \pmod{p^n}$$

is obtained by substituting $x = x_0 + x_1p + x_2p^2 + \ldots$. Then x_0 is the solution of $a\,x \equiv b \pmod{p}$, and $x_1, x_2, x_3, \ldots$ are obtained consecutively by equating co-efficients of p, p^2, p^3, etc., each giving a congruence $\pmod{p}$.

Congruences of higher degree may be treated similarly. Thus, the congruence

$$x^2 + 3x + 5 \equiv 0 \pmod{5}$$

has two solutions, $x \equiv 0$ and $x \equiv 2 \pmod{5}$.
Corresponding to each solution $\pmod{5}$ there is a unique solution of the same congruence $\pmod{125}$.
Corresponding to $x \equiv 0 \pmod{5}$, put

$$x \equiv a\,5 + b\,25 \pmod{125}.$$

The congruence gives

$$25a^2 + 15a + 75b + 5 \equiv 0 \pmod{125}.$$

Taking this congruence $\pmod{25}$,

$$15a + 5 \equiv 0 \pmod{25},$$
$$3a + 1 \equiv 0 \pmod{5},$$
$$a \equiv 3 \pmod{5}.$$

Taking $a = 3$, the congruence (mod 125) gives

$$225 + 45 + 75b + 5 \equiv 0 \pmod{125}$$
$$275 + 75b \equiv 0 \pmod{125}.$$

Divided by 25

$$11 + 3b \equiv 0 \pmod 5$$

whence
$$b \equiv 3 \pmod 5.$$

$x = 90$ is thus a solution of the congruence

$$x^2 + 3x + 5 \equiv 0 \pmod{125}.$$

Similarly the solution (mod 125) corresponding to $x \equiv 2 \pmod 5$ can be obtained as $x \equiv 32$. These are the only two solutions of the congruence.

An entirely different circumstance obtains from the congruence:

$$x^2 + 3x + 6 \equiv 0 \pmod{125}.$$

The corresponding congruence (mod 5) can be expressed

$$x^2 - 2x + 1 = (x - 1)^2 \equiv 0 \pmod 5,$$

which is a quadratic with equal roots. Any solution must be congruent to 1 (mod 5).

Putting $x = 1 + 5y$ we obtain

$$1 + 10y + 25y^2 + 3 + 15y + 6 = 10 + 25y + 25y^2 \equiv 0 \pmod{125}$$
$$2 + 5y + 5y^2 \equiv 0 \pmod{25}.$$

But of this congruence there is clearly no solution.

On the other hand the congruence

$$x^2 + 3x - 29 \equiv 0 \pmod{125}$$

leads to the same congruence (mod 5), but in this case the substitution $x = 1 + 5a$ gives

$$-25 + 25a + 25a^2 \equiv 0 \pmod{125},$$
$$-1 + a + a^2 \equiv 0 \pmod 5.$$

The only solution of this congruence is $a \equiv 2 \pmod 5$, but this gives five solutions of the original congruence, namely

$$x \equiv 11,\ 36,\ 61,\ 86,\ 111 \pmod{125}.$$

Such congruences as $x^2 + 3x + 6 \equiv 0$, and $x^2 + 3x - 29 \equiv 0$, (mod 125), are called singular congruences. They correspond closely to the linear congruences obtained from attempted division by a singular residue.

In the general case, if the congruence $f(x) \equiv 0 \pmod p$, has a solution or root $x \equiv \alpha \pmod p$, then $f(x) \equiv g(x)(x - \alpha)$, (mod p), for some polynomial $g(x)$. If also $f(x) \equiv h(x)(x - \alpha)^2$

(mod p), then $x \equiv a$ is called a repeated root of the congruence $f(x) \equiv 0 \pmod{p}$.

If a is a root but not a repeated root of the congruence $f(x) \equiv 0 \pmod{p}$, then there is a unique root a_1 of $f(x) \equiv 0 \pmod{p^n}$, such that $a_1 \equiv a \pmod{p}$.

But if a is a repeated root of the congruence $f(x) \equiv 0$ (mod p), then either $f(x) \equiv 0 \pmod{p^n}$, will have no root, or it will have at least p roots congruent to $a \pmod{p}$.

COMPOSITE MODULI

Before considering the general modulus we examine first congruences to modulus $p\ q$, where p and q are distinct primes.

Firstly for any given pair of numbers A, B, there is a unique residue x, (mod pq) such that $x \equiv A \pmod{p}$, and $x \equiv B \pmod{q}$. For Euclid's Algorithm gives

$$Pp + Qq = 1$$

and

$$A\ Qq \equiv A \pmod{p},$$
$$B\ Pp \equiv B \pmod{q},$$

so that

$$x = A\ Qq + B\ Pp$$

gives $x \equiv A \pmod{p}$ and $x \equiv B \pmod{q}$.

This completely solves the problem of congruences (mod pq), by reducing it to the corresponding problem with prime modulus.

Thus to solve the congruence

$$f(x) \equiv 0 \pmod{pq}$$

we consider the subsidiary equations

$$f(x) \equiv 0 \pmod{p};\ f(x) \equiv 0 \pmod{q}.$$

Then if A and B are any pair of roots of these two equations, these define a unique residue x (mod pq) such that $x \equiv A \pmod{p}$, and $x \equiv B \pmod{q}$. This gives

$$f(x) \equiv f(A) \equiv 0 \pmod{p},$$
$$f(x) \equiv F(B) \equiv 0 \pmod{q},$$

and thus

$$f(x) \equiv 0 \pmod{pq}.$$

But notice that a quadratic congruence (mod pq), may have four roots, rather than two. For if $f(x)$ is a quadratic either of the roots of $f(x) \equiv 0 \pmod{p}$, may be combined with either of the roots of $f(x) \equiv 0 \pmod{q}$, to give four combinations, any of which gives a root of $f(x) \equiv 0 \pmod{pq}$.

Thus $x^2 \equiv 1 \pmod 5$ gives $x \equiv 1$ or $4 \pmod 5$
and $x^2 \equiv 1 \pmod 7$ gives $x \equiv 1$ or $6 \pmod 7$.
Then $x \equiv 1 \pmod 5$, $x \equiv 1 \pmod 7$ gives $x \equiv 1 \pmod{35}$,
$x \equiv 1 \pmod 5$, $x \equiv 6 \pmod 7$ gives $x \equiv 6 \pmod{35}$,
$x \equiv 4 \pmod 5$, $x \equiv 1 \pmod 7$ gives $x \equiv 29 \pmod{35}$,
$x \equiv 4 \pmod 5$, $x \equiv 6 \pmod 7$ gives $x \equiv 34 \pmod{35}$.
These are the four roots of $x^2 \equiv 1 \pmod{35}$.

In a similar way the congruence $f(x) \equiv 0 \pmod{p^n\, q^m}$, can be made to depend on the separate congruences $f(x) \equiv 0 \pmod{p^n}$, and $f(x) \equiv 0 \pmod{q^m}$.

In a like manner the congruence $f(x) \equiv 0 \pmod{p_1^{n_1}\, p_2^{n_2} \ldots p_r^{n_r}}$ can be made to depend on the set of congruences $f(x) \equiv 0 \pmod{p_r^{n_r}}$, $r = 1, 2, \ldots$ i. This is the most general integral modulus.

CHAPTER V

POLYNOMIALS

THE symbol x is used very extensively in mathematics. The exact significance of the letter, however, varies subtly with the context. In the early stages of algebra x denotes some unknown number, which it is the object of the work to discover. Later, with the ideas of proportion, variation and of functions, there is a change of meaning which passes almost unnoticed.

The letter no longer signifies some specific number, but has more generality. It can range over all numbers, or over all numbers of a given set. The "unknown" x is replaced by the "variable" x.

But unknowns and variables obey exactly the same laws and it is not always necessary to have a clear idea which one is using. Sometimes one passes from unknown to variable without noticing. Thus it may be required to solve the equation $x^3 + 3x + 2 = 0$. In this, x is definitely an unknown. But quite a convenient method of solving is to plot the graph of the equation $y = x^3$ and find the intersection with the line $y + 3x + 2 = 0$. In the graph the symbol x has ceased to be an unknown and has become a variable.

The laws of algebra are not concerned, however, in whether x represents an unknown, a variable or a constant. A further important generalization is here introduced. The letter x can denote an "indefinable". That is to say, it is just a symbol concerning which nothing is assumed, except that it obeys the fundamental laws of algebra. This implies that it can be added, subtracted, multiplied and possibly divided by numbers or by other indefinables, and that these operations of addition and multiplication are commutative and associative, and that multiplication is distributive with respect to addition.

Unknowns have all these properties, and hence the indefinable can be assumed at any stage to be an unknown, if we so wish it. Variables also have these properties, so that any statement about an indefinable includes the same statement about a variable. But a statement about an indefinable is more general, since it is not assumed that the indefinable represents a number of the system at all. In fact, in many instances indefinables are used to augment the number system.

But let us first examine the number systems themselves. Several number systems have been defined, the signed integers, the rationals, real numbers and the various modular systems.

In all these systems the sum, difference and product of two numbers of the system are also numbers of the system, and the operations of addition and multiplication obey the five associative, commutative and distributive laws. A system satisfying these conditions is called a *commutative ring*.

A commutative ring which has the added property that if $x\,y = 0$ then either $x = 0$ or $y = 0$ is called an *integral domain*. It follows from this that if $a\,x = a\,y$ and $a \neq 0$, then $a\,(x - y) = 0$ and $x = y$. Clearly the integers, the rationals and real numbers form integral domains, as do the residues to a prime modulus. But the residues to a modulus which is not prime do not form an integral domain.

An integral domain in which division is possible by any number except zero is called a *field*. The integers do not form a field, but the rationals, the real numbers, and the residues to a prime modulus each form a field.

Now take any integral domain D and an indefinable x. The operations of addition and multiplication lead to a new integral domain of which the general member is of the form

$$a_0 + a_1 x + a_2\,x^2 + \ldots + a_n\,x^n$$

where the coefficients $a_0, a_1, \ldots, a_n$ are numbers of the domain D.

Such an expression is called a *polynomial in the indefinable x over the domain D*. The *degree* of the polynomial is n.

If the domain D is a field which we will denote by F, then the expression is a polynomial over the field F.

Polynomials over a field have many of the properties of integers. Firstly Euclid's Algorithm is applicable.

Let $f(x)$ and $g(x)$ be two polynomials over a field, and suppose that the degree of $f(x)$ is greater than or equal to that of $g(x)$. Divide $f(x)$ by $g(x)$ until a remainder $r_1(x)$ is obtained such that the degree of $r_1(x)$ is less than that of $g(x)$. Then divide $g(x)$ by $r_1(x)$ until a remainder $r_2(x)$ is obtained of still lower degree. Proceeding similarly, exactly as for Euclid's Algorithm for integers, the last non-zero remainder is the highest common factor, or H.C.F., of $f(x)$ and $g(x)$. Denoting this by $h(x)$, the same procedure as with the integers gives polynomials $F(x)$ and $G(x)$ such that

$$F(x)\, f(x) + G(x)\, g(x) = h(x).$$

In particular, if $f(x)$ and $g(x)$ have no common factor, then

$$F(x)\, f(x) + G(x)\, g(x) = 1.$$

Thus for $f(x) = x^3 + 3x + 1$, $g(x) = x^2 - x + 2$

$$f(x) - (x + 1)\, g(x) = 2x - 1$$

$$g(x) - (\tfrac{1}{2}x - \tfrac{1}{4})\,(2x - 1) = 1\tfrac{3}{4}$$

Thus
$$\begin{aligned} 7 &= 4\, g(x) - (2x - 1)\,(2x - 1) \\ &= 4\, g(x) - (2x - 1)\,(f(x) - (x + 1)\, g(x)) \\ &= (2\, x^2 + x + 3)\, g(x) - (2x - 1)\, f(x). \end{aligned}$$

So that
$$\frac{1}{7}\,(2x^2 + x + 3)\, g(x) - \frac{1}{7}\,(2x - 1)\, f(x) = 1.$$

Notice that although the coefficients in $f(x)$ and $g(x)$ are integers, yet fractions must be introduced in the coefficients of $F(x)$ and $G(x)$. For Euclid's Algorithm to be applicable to polynomials, the coefficients must be taken over a field; an integral domain will not suffice.

Just as integers can be prime or composite, so can polynomials. But whether a polynomial is regarded as prime or composite depends upon the field over which the coefficients are taken.

Thus, over the real numbers, the quadratic $x^2 + 4x + 1$ is composite, being factorized as

$$x^2 + 4x + 1 = (x + 2 + \sqrt{3})\,(x + 2 - \sqrt{3}).$$

But if the field is that of the rationals, then $(x + 2 \pm \sqrt{3})$ are not polynomials over the field, and the quadratic $x^2 + 4x + 1$ is not composite but is prime.

Following exactly the proof given for integers, Euclid's Algorithm can be used to show that every polynomial over a given field can be expressed in one and only one way as a product of prime polynomials. For this purpose, however, a numerical factor belonging to the field is ignored. Thus

$$x^2 - 1 = (x + 1)(x - 1) = (2x + 2)(\tfrac{1}{2}x - \tfrac{1}{2})$$

would be regarded as the same factorization. Uniqueness can be restored by the assumption that the coefficient of the highest power of x in each prime factor is unity.

Though the coefficients are taken from a field, the polynomials do not form a field, but an integral domain.

Fields can be formed from this integral domain by various devices.

Firstly, that method can be employed by means of which rationals were defined in terms of integers. Thus, consider pairs of polynomials $f(x)$, $g(x)$, the pair being written for convenience in the form $f(x)/g(x)$. It is assumed that $g(x) \neq 0$. A definition of equality is taken so that $f(x)/g(x) = h(x)/k(x)$ if and only if $f(x)\,k(x) = g(x)\,h(x)$. If addition and multiplication are defined exactly as for rationals, the pairs of polynomials $f(x)/g(x)$, which are called *rational fractions*, form a field.

If the denominator of a rational fraction is composite, the rational fraction can be put into partial fractions in the same way as for rationals, Euclid's Algorithm being used to effect this.

Thus if $f(x)$, $g(x)$ are co-prime, i.e. if they have no common factor, then $F(x)$ and $G(x)$ can be found so that

$$F(x)\,f(x) + G(x)\,g(x) = 1.$$

Then $\quad F(x)/g(x) + G(x)/f(x) = 1/f(x)\,g(x),$

and $\quad k(x)/(f(x)\,g(x)) = k(x)\,F(x)/g(x) + k(x)\,G(x)/f(x).$

Secondly, the integral domain of polynomials over a field can be extended to form a field by the use of infinite series.

If a_1 differs from zero, the polynomial

$$a_0 + a_1 x + a_2 x^2 + \ldots + a_n x^n$$

can be formally divided into unity to give as a quotient an infinite series of ascending powers of x beginning with

$$\frac{1}{a_0} - \frac{a_1}{a_0^2} x + \frac{a_1^2 - a_0 a_2}{a_0^3} x^2 - \ldots$$

No question of convergence arises. The system of polynomials in x is just extended to include such infinite series.

The system is not yet a field, as the element x has no reciprocal. But the system of polynomials and series of the form

$$x^p\ (a_0 + a_1\ x + a_2\ x^2 + \ldots)$$

where p is a positive or negative integer does form a field.

Thirdly, fields were obtained from the natural integers by taking the system of residues to a prime modulus. The same procedure is applicable with polynomials over a field.

If $f(x)$ is a prime polynomial of degree n over a field F, then $P(x)$ and $Q(x)$ are said to be congruent to modulus $f(x)$ if $P(x) - Q(x)$ is divisible by $f(x)$. We write

$$P(x) \equiv Q(x), \text{ (mod } f(x)).$$

Clearly there is a unique polynomial $r(x)$ of degree less than n, such that

$$P(x) \equiv r(x) \text{ (mod } f(x)).$$

Then $r(x)$ is called a residue (mod $f(x)$).

The product of two residues is congruent to a residue (mod $f(x)$). Thus the residues form an integral domain.

Further, apart from zero, every residue has a reciprocal, for since the degree of $r(x)$ is less than that of $f(x)$ and $f(x)$ has no factor, then $f(x)$ and $r(x)$ are co-prime. Euclid's Algorithm gives

$$R(x)\ r(x) + F(x)\ f(x) = 1.$$

Thus $$R(x)\ r(x) \equiv 1 \text{ (mod } f(x)),$$

and $R(x)$ is the reciprocal of $r(x)$.

Hence the residues of a system of polynomials over a field, when the modulus is a prime polynomial $f(x)$, constitute a field.

This procedure will be adopted in the next chapter for the definition of complex numbers, and also to construct various fields of algebraic numbers by extensions of the rationals.

But attention is now drawn especially to the generality of the whole theory, in that the identical procedures employed for the natural integers fit perfectly into the domain of polynomials.

CHAPTER VI

COMPLEX NUMBERS AND ALGEBRAIC FIELDS

SINCE the square of every real number is positive, the equation $x^2 + 1 = 0$ can have no root over the real numbers. Let i be an indefinable. Then the polynomial $(i^2 + 1)$ is irreducible over the real numbers. Hence, as was shown in the last chapter, the system of residues to modulus $(i^2 + 1)$ with coefficients over the real numbers forms a field.

These residues are the polynomials of degree less than two in i, that is, the expressions of the form $a + bi$, where a and b are real numbers.

Expressions of the form $a + bi$, with a and b real numbers, treated as residues to modulus $(i^2 + 1)$ are called *complex numbers.*

The complex numbers have a unique property unshared by any other field, apart from the sub-field of all the algebraic complex numbers. Every polynomial of degree greater than one over the complex numbers is reducible. Thus a polynomial of degree n is expressible as a product of n linear factors. An equation of degree n, $f(x) = 0$ is expressible in the form

$$(a_1 x + b_1)(a_2 x + b_2) \ldots (a_n x + b_n) = 0$$

which has the n roots,

$$x = -b_1/a_1, \; -b_2/a_2, \ldots, -b_n/a_n.$$

This is not the place to extoll the virtues of complex numbers, which have widespread application to almost every branch of mathematics. Suffice it to say that because of their unique property that every polynomial is reducible into linear factors they bring a measure of perfection and completion to each subject to which they are applied.

But because of this very perfection, our methods have no further application here, and no extension of the complex number field is possible. But the real numbers themselves constitute a very vast extension of the rationals, and it is in this region between the rationals and the reals that the procedure provides a powerful method of exploration.

Let a be a rational number which is not an exact square. Then $(\xi^2 - a)$ considered as a polynomial in ξ, is irreducible over the field of rationals. Hence if ξ is an indefinable, the system of residues (mod $(\xi^2 - a)$), i.e. expressions of the form $b\,\xi + c$, where b and c are rationals, form a field.

For example, taking $a = 7$, the expressions of the form $b + c\sqrt{7}$ form a field and the quotient of two expressions of this form can be expressed in the same form. Thus

$$(6 + 2\sqrt{7}) \div (2 + 3\sqrt{7}) = \frac{6 + 2\sqrt{7}}{2 + 3\sqrt{7}} \cdot \frac{3\sqrt{7} - 2}{3\sqrt{7} - 2}$$
$$= \frac{42 - 12 + 18\sqrt{7} - 4\sqrt{7}}{63 - 4} = \frac{30}{59} + \frac{14\sqrt{7}}{59}$$

Such expressions are dealt with in most elementary algebra books under the title of *surds*. The interesting fact is that the introduction of a particular surd leads to an extension of the rational number system which retains the property of being a field. It is usual to denote the field of rationals by R. Then the field of expressions of the form $b + c\xi$ where $\xi^2 = a$ is denoted by $R(\xi)$ or $R(\sqrt{a})$. It is called the field obtained by *adjoining* ξ to the rationals.

More generally if $f(\xi)$ is any polynomial of degree n which is irreducible over the rationals, the system of residues (mod $f(\xi)$), i.e. expressions of the form $a + b\,\xi + c\,\xi^2 + \ldots + k\,\xi^{n-1}$ where the coefficients $a, b, c, \ldots, k$ are rational, form a field, denoted by $R(\xi)$. The indefinable ξ may be identified with any of the real or complex roots of the equation $f(x) = 0$.

For example if $\xi^3 - \xi + 2 = 0$, then the quotient of say $\xi^2 + 1$ and $\xi^2 + \xi - 1$ can be expressed in the form $a + b\,\xi + c\,\xi^2$, with a, b, c rational. The procedure is as follows. Go through Euclid's Algorithm for the expressions $\xi^3 - \xi + 2$ and $\xi^2 + \xi - 1$. Thus

$$\xi^3 - \xi + 2 - (\xi - 1)(\xi^2 + \xi - 1) = \xi + 1,$$
$$\xi^2 + \xi - 1 - \xi(\xi + 1) = -1.$$

Then

$$\begin{aligned} 1 &= \xi(\xi + 1) - (\xi^2 + \xi - 1) \\ &= \xi(\xi^3 - \xi + 2) - (\xi^2 - \xi)(\xi^2 + \xi - 1) - (\xi^2 + \xi - 1) \\ &= \xi(\xi^3 - \xi + 2) - (\xi^2 - \xi + 1)(\xi^2 + \xi - 1). \end{aligned}$$

Hence since $\xi^3 - \xi + 2 = 0,$

$$(\xi^2 - \xi + 1)(\xi^2 + \xi - 1) = -1$$

and $1/(\xi^2 + \xi - 1) = -\xi^2 + \xi - 1$

so that

$$\begin{aligned} \frac{\xi^2 + 1}{\xi^2 + \xi - 1} &= (\xi^2 + 1)(-\xi^2 + \xi - 1) \\ &= -\xi^4 + \xi^3 - 2\xi^2 + \xi - 1 \\ &= -(\xi - 1)(\xi^3 - \xi + 2) - 3\xi^2 + 4\xi - 3 \\ &= \underline{\underline{-3\xi^2 + 4\xi - 3.}} \end{aligned}$$

If $f(x)$ is an irreducible polynomial over the rationals, it will be reducible over the field $R(\xi)$ where $f(\xi) = 0$, having the factor $(x - \xi)$. If $f(x)$ is of degree n, then the removal of the factor $(x - \xi)$ leaves a polynomial of degree $n - 1$, which in general will be irreducible over $R(\xi)$. Except in the case of a quadratic, the determination of one root of $f(x) = 0$ does not allow the others to be deduced.

The polynomial $f(x)$ is not the only one, which, being irreducible over the rationals becomes reducible over $R(\xi)$. It is well known that the quadratic equation $a\,x^2 + b\,x + c = 0$ can be solved in terms of rationals and the single irrational $\sqrt{(b^2 - 4\,a\,c)}$. Hence if $\xi^2 = a$ and $b^2 - 4\,a\,c = k^2\,a$, with k rational, then the polynomial $a\,x^2 + b\,x + c$ is irreducible over R but reducible over $R(\xi)$.

More generally, if $f(\xi) = 0$ defines a field $R(\xi)$, and $g(\xi)$ is any polynomial of lower degree in ξ, then $\eta = g(\xi)$ is not rational, but will satisfy an equation of the same degree as $f(\xi)$, say $F(\eta) = 0$. In general $F(x)$ will be irreducible in R but reducible in $R(\xi)$. In this case $R(\eta)$ is identical with $R(\xi)$. It might happen in special cases, however, that η satisfies an equation of lower degree, when $R(\eta)$ would be contained in $R(\xi)$, that is to say η could be expressed in terms of ξ but not ξ in terms of η.

As an example, if

$$f(\xi) \equiv \xi^3 - \xi - 1 = 0$$

and

$$\eta = 1 + \xi + \xi^2$$

then

$$\eta^2 = 1 + 2\xi + 3\xi^2 + 2\xi^3 + \xi^4 = 3 + 5\xi + 4\xi^2,$$
$$\eta^3 = 3 + 8\xi + 12\xi^2 + 9\xi^3 + 4\xi^4 = 12 + 21\xi + 16\xi^2.$$

Then $$\eta^3 - 4\eta^2 = \xi,\ \eta^2 - 4\eta = \xi - 1,$$

and $$\eta^3 - 5\eta^2 + 4\eta - 1 = 0.$$

In this case ξ can be expressed in terms of η as

$$\xi = \eta^3 - 4\eta^2.$$

The special case arises in the following example. Suppose that

$$f(\xi) \equiv \xi^4 - 2 = 0,$$

and $\eta = \xi^2 + 1.$

Then $\eta^2 = \xi^4 + 2\xi^2 + 1 = 2\xi^2 + 3 = 2\eta + 1,$

and $\eta^2 - 2\eta - 1 = 0.$

Thus it satisfies a quadratic equation instead of a quartic. Though η has been expressed in terms of ξ, yet ξ cannot be expressed rationally in terms of η.

The equation $x^4 - 2 = 0$ is a special type of equation, for it can be solved by solving consecutively two quadratic equations, e.g.

$$y^2 - 2y - 1 = 0,$$

and $$x^2 + 1 = y.$$

It is only in such exceptional cases that the degree of $F(\eta)$ can be less than the degree of $f(\xi)$. More light on these equations is shed by the Galois Theory of Equations, which will be mentioned later.

So far we have extended the field of rationals by adjoining one irrational ξ which satisfies an equation of degree n, namely $f(\xi) = 0$.

But the field $R(\xi)$ can be treated in exactly the same way. Suppose that η satisfies an equation $g(\eta) = 0$ of degree m with coefficients in $R(\xi)$ and irreducible over $R(\xi)$. Then the polynomials of degree less than m in η with coefficients in $R(\xi)$, or what amounts to the same thing, the polynomials in ξ and η of degree less than n in ξ and less than m in η, form a field denoted by $R(\xi, \eta)$.

The interesting fact is that we can always find a single equation $F(x) = 0$ which is of degree mn such that if ζ satisfies $F(\zeta) = 0$, then both ξ and η can be expressed in terms of ζ and hence $R(\zeta)$ is the same field as $R(\xi, \eta)$.

As an example let $f(\xi) \equiv \xi^2 - 2 = 0$, and $g(\eta) \equiv \eta^3 - 3 = 0$.
Put $Z = \xi + \eta$
then
$$Z - \xi = \eta$$
and
$$\eta^3 = 3 = Z^3 - 3\,Z^2\,\xi + 3\,Z\,\xi^2 - \xi^3$$
$$= Z^2 + 6\,Z - \xi(3\,Z^2 + 2)$$
$$Z^3 + 6\,Z - 3 = \xi\,(3\,Z^2 + 2)$$
$$(Z^3 + 6\,Z - 3)^2 = \xi^2\,(3\,Z^2 + 2)^2 = 2(3\,Z^2 + 2)^2$$
or $F(Z) = Z^6 - 6\,Z^4 - 6\,Z^3 + 12\,Z^2 - 36\,Z + 1 = 0$.
If $F(\zeta) = 0$, then
$$\xi = (\zeta^3 + 6\zeta - 3)/(3\,\zeta^2 + 2)$$
which can be expressed as a polynomial in ξ of degree less than 6, by the use of Euclid's Algorithm, as has been shown. Similarly η can be expressed in terms of ζ, and $R(\zeta)$ is the same as $R(\xi, \eta)$.

Notice that this sextic equation $F(z) = 0$ is again a special kind of equation, for we can solve it by solving separately a cubic and a quadratic, and this is not true of the general sextic.

But if we work at the problem from the other end, if we are given a sextic equation, for example, and want to find out whether it is a special type of equation which can be solved by means of an auxiliary cubic and an auxiliary quadratic, then much more difficulty arises. It is of such problems that the Galois Theory is concerned.

Suppose that we are given a sextic equation and wish to make it completely resolvable into linear factors by adjoining irrationals to the rational field. To find the first root of the sextic, it is necessary to adjoin to the field a root of the sextic. On removing the linear factor corresponding to this known root, there remains a quintic equation which in general is irreducible. A further root of this quintic must be adjoined, and on removal of the corresponding linear factor, a root of the remaining quartic, and consecutively a cubic and a quadratic.

It has been mentioned that the adjoining of a pair of roots of equations of respective degrees *m* and *n* to the rationals is equivalent to the adjoining of one root of an equation of degree *mn*. Hence the adjoining of all the roots of a sextic to the rationals is equivalent to the adjoining of one root of an equation of degree 6 . 5 . 4 . 3 . 2 = 720. There is an equation of degree 720 such that all the roots of the sextic can be expressed rationally in terms of any one root of the 720th degree equation. Such an equation is called the *complete Galois resolvent*. In general the equation is irreducible, but if the sextic is a special type of equation the complete Galois resolvent may be reducible.

Some study of the problems which arise is given in the chapter on Galois' theory of equations.

CHAPTER VII

ALGEBRAIC INTEGERS, IDEALS AND p-ADIC NUMBERS

CONSIDERED as a real number, 5 is prime. In the complex numbers, however, 5 can be expressed as a product in the form

$$5 = (2 + i)(2 - i).$$

It is convenient to regard $(2 + i)$ and $(2 - i)$ as complex integers. In general, over the complex numbers $(m + ni)$ is said to be an integer if m and n are real integers.

At first sight it might appear that the structure of complex integers differed from that of reals, in that factorization of 5 could be expressed in another way as

$$5 = (1 + 2i)(1 - 2i).$$

The same difficulty, however, occurs among the reals, for 21 could be factored in two ways as

$$21 = 3 \times 7 = (-3)(-7).$$

In practice we ignore the factor -1, and regard these two factorizations as the same. The numbers 1 and -1 are called units, and units are ignored in the factorization. Thus 3 and -3, of which one can be obtained from the other by multiplication by a unit, are not discriminated.

In complex numbers there are four units 1, -1, i and $-i$. A unit is defined as an integer of which the reciprocal exists and is also an integer. Clearly, if $n = pq$ and u and v are units such that $uv = 1$, then also $n = (up)(vq)$. Hence as factors of n, p and up are regarded as the same factor. Thus since $(1 + 2i) = i(2 - i)$, the factors $(1 + 2i)$ and $(2 - i)$ are equivalent, and similarly $(2 + i)$ and $(1 - 2i)$.

With this convention factorization is unique both for real and complex numbers. The proof of the uniqueness for complex integers depends once again on Euclid's Algorithm.

The *norm* of a complex number $(a + ib)$ is defined as $(a^2 + b^2)$. In applying Euclid's Algorithm to find the highest common factor of two complex numbers w and z, we divide the one with the larger norm, say w, by the other so as to obtain a remainder whose norm is less than the norm of z. With this difference the procedure is exactly the same as for real integers.

Real primes of the form $n = 4r - 1$ are also prime over the complex numbers. All other real primes, i.e. 2 and odd primes of the form $n = 4r + 1$ are composite over the complex numbers. Thus

$2 = (1 + i)(1 - i)$, $5 = (2 + i)(2 - i)$, $13 = (3 + 2i)(3 - 2i)$.

The theory of numbers extends naturally to complex numbers. If $p = 4r - 1$ is a real prime, then the system of complex residues (mod p) are p^2 in number, being of the form $m + in$, where m and n are real residues (mod p). Omitting the zero residue, these form a field of order $p^2 - 1$. Every non-zero residue satisfies the congruence

$$x^{p^2-1} \equiv 1 \pmod{p}.$$

There are quadratic residues which satisfy

$$x^{\frac{1}{2}(p^2-1)} \equiv 1 \pmod{p}$$

and also quadratic non-residues which satisfy

$$x^{\frac{1}{2}(p^2-1)} \equiv -1 \pmod{p}.$$

If α is a complex number whose norm is q, a real prime number of the form $4r + 1$, then there exists a system of complex residues (mod α), which are q in number. These satisfy

$$x^{q-1} \equiv 1 \pmod{\alpha}.$$

The quadratic residues satisfy

$$x^{\frac{1}{2}(q-1)} \equiv 1 \pmod{\alpha},$$

and the quadratic non-residues

$$x^{\frac{1}{2}(q-1)} \equiv -1 \pmod{\alpha}.$$

The law of quadratic reciprocity can be extended to these complex residues.

The concept of integers can be extended also to the algebraic fields. For example in $R(\sqrt{2})$ numbers of the form $m + n\sqrt{2}$, where m and n are cardinal integers, are called algebraic integers.

The system of units is more extensive in $R(\sqrt{2})$. Since $(\sqrt{2} - 1)(\sqrt{2} + 1) = 1$, both $(\sqrt{2} - 1)$ and $(\sqrt{2} + 1)$ are units as are also $(\sqrt{2} - 1)^n$ and $(\sqrt{2} + 1)^n$. There are thus an infinity of units, and if we had not decided to ignore units in factorization, any number could be factored in an infinite number of ways. Thus

$$2 = (2 + 2\sqrt{2})(\sqrt{2} - 1) = (6 + 4\sqrt{2})(3 - 2\sqrt{2})$$
$$= (10\sqrt{2} + 14)(5\sqrt{2} - 7).$$

Ignoring the units, however, these factorizations are equivalent.

As with real and complex numbers, systems of residues may be defined in $R(\sqrt{2})$. Factorization is unique and Euclid's Algorithm is applicable. The norm of the integer $a + b\sqrt{2}$ is defined as $a^2 - 2b^2$, or $2b^2 - a^2$, whichever is positive. The procedure is then exactly as with complex integers.

In $R(\sqrt{5})$ a new circumstance arises, for

$$(\sqrt{5} - 1)(\sqrt{5} + 1) = 4 = 2 \times 2,$$

and two distinct factorizations of 4 arise. In this case, however, it is convenient to regard $(\sqrt{5} + 1)$ as being divisible by 2, that is to say, $\frac{1}{2}(\sqrt{5} + 1)$ is regarded as an integer.

This is a fitting occasion to discuss the proper definition of an algebraic integer. The definition should extend to domains to which the roots of equations have been adjoined other than the case $x^2 = n$.

If $f(x) = 0$ is an equation with integral coefficients such that the coefficient of the highest power of x is unity, then an algebraic number ξ satisfying $f(\xi) = 0$ is called an algebraic integer.

The number $\xi = \frac{1}{2}(\sqrt{5} + 1)$ satisfies

$$x^2 - x - 1 = 0$$

and is thus an integer.

Since $\frac{1}{2}(\sqrt{5} + 1) \times \frac{1}{2}(\sqrt{5} - 1) = 1$, each of these integers is a unit. $(\sqrt{5} + 1)$ and 2 are therefore equivalent factors of 4.

With these conventions, Euclid's Algorithm is found to be applicable in $R(\sqrt{5})$, factorization becomes once again unique, and the theory of congruences and residues are valid.

Such is not the case, however, in $R(\sqrt{-5})$. The two factorizations of 6, namely

$$6 = (\sqrt{-5} + 1)(1 - \sqrt{-5}) = 3 \times 2$$

cannot be reconciled by any intepretation of the term algebraic integer.

Euclid's Algorithm fails in $R(\sqrt{-5})$ because, on dividing an integer w by another integer z, it is sometimes impossible to get a remainder whose norm is less than the norm of z. Thus on dividing $\sqrt{-5}+1$ by 2, there can be no smaller remainder than $\pm\sqrt{-5}\pm 1$, of which the norm is 6, greater than the norm of 2, which is 4.

The consequences of Euclid's Algorithm no longer apply, and hence is made possible the two separate factorizations of 6.

The theory of congruences and residues breaks down, for the residues may not form a field. Thus to modulus 3, the number $\sqrt{-5}+1$ would be a singular residue, for $(\sqrt{-5}+1)\,(1-\sqrt{-5}) \equiv 0 \pmod{3}$. The modulus 3 behaves more like a composite than like a prime residue.

To restore uniqueness of factorization and to make possible a theory of congruences and residues, the concept of *ideal numbers* or *ideals* is introduced. This will be explained first with reference to the rational integers.

The set of all the rational integers is closed to the operations of addition, subtraction and multiplication. This set forms the unit ideal denoted by [1].

The set contains subsets which are closed to addition, subtraction and multiplication. Suppose that the number 3 is the smallest integer included. Since it is closed to addition it contains $3+3=6$, $6+3=9$, and every multiple of 3. It cannot contain any other integer since, being closed to subtraction, it would also contain a number less than 3. Thus if 35 were a member, then also would be $35-33=2$ and $3-2=1$. Hence the set contains all multiples of 3 and is denoted by [3]. These sets are called ideals. The product of two ideals is the set of numbers which are the product of two numbers, one from each ideal, together with all numbers that can be formed from these by addition and subtraction. The product is thus an ideal.

Thus the product of the ideals $[2]\times[3]$ contains the numbers $2\times 3=6$, $4\times 3=12$, $2\times 9=18$, etc., and forms the ideal [6].

An ideal which contains all the multiples of a given integer n and no other numbers is called a *principal ideal*, and is denoted by $[n]$.

As regards multiplication, principal ideals are simply isomorphic with the numbers to which they correspond. Over the rationals, all ideals are principal ideals, and nothing new arises from their introduction.

Over the field $R(\sqrt{-5})$, however, there exist also ideals which are not principal, and these form an extension of the system of integers in which the uniqueness of prime factorization is restored.

If α and β are any two integers in $R(\sqrt{-5})$, then the set of numbers

$$\alpha 2 + \beta(1 + \sqrt{-5})$$

is closed to addition and subtraction, and is closed to multiplication by any integer in $R(\sqrt{-5})$. It is called an ideal and is denoted by $[2, 1 + \sqrt{-5}]$. It will be represented by p.

Now p^2 contains the terms

$$4,\ 2 + 2\sqrt{-5},\ -4 + 2\sqrt{-5},$$

and hence by combination

$$(2 + 2\sqrt{-5}) - (-4 + 2\sqrt{-5}) = 6$$

and $$6 - 4 = 2.$$

Also there is $$(2 + 2\sqrt{-5}) - 2 = 2\sqrt{-5}.$$

It is clear that p^2 is equal to the principal ideal $[2]$.

Now put

$$q = [3,\ 1 + \sqrt{-5}],$$
$$r = [3,\ 1 - \sqrt{-5}].$$

Then $q\,r$ contains the terms

$$9,\ 3 + 3\sqrt{-5},\ 3 - 3\sqrt{-5},$$

and hence

$$(3 + 3\sqrt{-5}) + (3 - 3\sqrt{-5}) = 6,\ 9 - 6 = 3,$$

and $$(3 + 3\sqrt{-5}) - 3 = 3\sqrt{-5}.$$

Thus $q\,r$ is the principal ideal $[3]$.

In a similar manner

$$p\,q = [1 + \sqrt{-5}],\ p\,r = [1 - \sqrt{-5}].$$

The number 6, or rather the principal ideal $[6]$ can be expressed as a product of prime ideals as

$$[6] = p^2\,q\,r.$$

The two factorizations $6 = 3 \times 2 = (1 + \sqrt{-5})(1 - \sqrt{-5})$ are just different ways of coupling the prime ideals.

The factorization into prime ideals is unique. Thence the theory of residues can be restored, for instead of taking the system of residues to the modulus of a prime number, the residues are taken to the modulus of a prime ideal. These residues form a field, and the same development is possible as in the case of the rational integers. The theorem of quadratic reciprocity can be extended to the ideals in an algebraic field.

The case of $R(\sqrt{-5})$ has been specially considered, but the same procedure is valid for any algebraic field.

While considering ideals and prime ideals, a brief description of the p-adic numbers is not out of place. These form extensions of the rational numbers which in many ways resemble the extension of the rational numbers to the real numbers. But whereas the extension to real numbers is unique and complete, the extensions to p-adic numbers are diverse, and always incomplete, for irreducible equations always exist, and from these may be obtained further extensions.

The extension of rationals to real numbers depends on magnitude, which is a "valuation".

A function $\theta(a)$ of numbers a is called a valuation if it satisfies the equations

$$\theta(a\,b) = \theta(a)\,\theta(b),$$
$$\theta(a+b) \leqslant \theta(a) + \theta(b).$$

For real numbers we take $\theta(a)$ to be the absolute value of a, that is $+a$ or $-a$, whichever is positive. For complex numbers we take $\theta(a+ib) = \sqrt{(a^2+b^2)}$.

A real number can then be defined by a sequence of rationals, the valuation of the difference becoming smaller and approaching zero the further one proceeds along the sequence. An infinite decimal is a convenient representation.

This valuation, however, is not the only possible valuation for rationals. Let p be any prime number. Then put $\theta(p) = r < 1$, and $\theta(n) = 1$ if n is prime to p. The equations $\theta(a)\,\theta(b) = \theta(ab)$ and $\theta(a/b) = \theta(a)/\theta(b)$ complete the definition of the valuation for rationals.

Consider the case $p = 3$. An approximation to any rational, say $5/7$, can be obtained as an integer, the valuation of the difference being less than one. For this it is only necessary to find an integer x such that $x \equiv 5/7 \pmod 3$ or

$$7x \equiv 5 \pmod 3$$

which gives $$x \equiv 2 \pmod 3.$$

In order that the valuation of the difference shall be less than r, it is necessary to take the equation

$$7x \equiv 5 \pmod{3^2}$$

or

$$x = 2 \pmod{3^2}.$$

Similarly, for the evaluation of the difference to be less than r^n, the congruence

$$7x \equiv 5 \pmod{3^{n+1}}$$

must be taken. This gives

$$x = 2 + 2 \cdot 3^2 + 3^3 + 2 \cdot 3^4 + 3^6 + 2 \cdot 3^8 + 3^9 + 2 \cdot 3^{10} + 3^{12} + \ldots$$

Notice that the coefficients are repeated after a period of six, including zero coefficients. Compare with the representation of 5/7 as a recurring decimal where the digits are repeated after a period of six. The fact that the period is six in each case is not fortuitous, but depends on the fact that for any number n prime to 7, $n^6 - 1$ is divisible by 7.

In order to represent all rationals, since powers of 3 may appear in the denominator, it is necessary to take infinite series of ascending powers of 3 which begin with some power which is possibly negative, thus

$$3^\alpha (a_0 + a_1 3 + a_2 3^2 + a_3 3^3 + \ldots)$$

where α is a zero, positive or negative integer, and each a_i either 0, 1 or 2.

These series bear a marked resemblance to decimals, but are, as it were, the wrong way round. When extending to infinity they extend in the direction of the ascending powers instead of the descending powers. Just as with decimals, when the sequence is finite, or when the coefficients after a certain point are recurring, the series represents a rational number. The extension of finite and recurring decimals to the general infinite decimals gives the extension of the rationals to the real numbers.

The corresponding extension of the rationals to the general infinite series of the form

$$3^\alpha (a_0 + a_1 3 + a_2 3^2 + \ldots)$$

is the system of p-adic numbers corresponding to the prime number 3. These numbers form a field. Though they are both obtained as limits of the rationals, p-adic numbers and real numbers are quite different. There is no correspondence between them and they have quite different properties.

It has been shown that there is one and only one algebraic extension of the real numbers, namely that to the complex numbers. The p-adic numbers, however, are capable of almost indefinite algebraic extension.

The equation

$$f(x) = 0$$

has a root which is a p-adic number in ascending powers of p, corresponding to each solution a of the congruence

$$f(x) \equiv 0 \pmod{p},$$

provided that $(x - a)$ is not a repeated factor of $f(x) \pmod{p}$.

Many congruences

$$f(x) \equiv 0 \pmod{p}$$

exist which have no solution, and corresponding to these the equation $f(x) = 0$ has no solution in the field of p-adic numbers. Introducing an indefinable x and taking residues (mod $f(x)$), an algebraic extension field is obtained.

The p-adic numbers have been described for the case when p is a prime rational integer. The more interesting developments occur, however, in algebraic fields, when the prime integer p is replaced by a prime ideal, and the p-adic numbers are usually defined with reference to prime ideals rather than prime numbers.

CHAPTER VIII

GROUPS

TAKE a pack of cards. To make the examination simple, suppose that we take quite a small pack, say eight cards from the ace to the eight of hearts, and we will take them in order with the ace uppermost. Now proceed to shuffle the cards, and to examine what is happening, use a regular method of shuffling them. Take the top ace, place the 2 on top, the 3 below, the 4 on top, the 5 below and so on until the eight cards are used. The order of the cards is then

$$8, 6, 4, 2, 1, 3, 5, 7.$$

The order is a little more mixed than it was at the start, but the regularity of the order is still apparent. Shuffle the cards again, then, by the same method. The order becomes

$$7, 3, 2, 6, 8, 4, 1, 5.$$

A further shuffle of the same kind gives

$$5, 4, 6, 3, 7, 2, 8, 1.$$

This third shuffle was less effective in mixing the cards, for the order seems rather more regular than after the second shuffle. A fourth shuffle brings the cards back to their original position,

$$1, 2, 3, 4, 5, 6, 7, 8.$$

Now the total number of ways of shuffling a pack of eight cards is $8 \times 7 \times 6 \times 5 \times 4 \times 3 \times 2 \times 1 = 40{,}320$. So it appears that this shuffling procedure is really rather ineffective, since its repetition will give only four out of forty thousand different shuffles. Is there a more effective shuffle, or will every shuffle come back to the identical shuffle if repeated a few times?

We will denote the above shuffle by S, and examine it more closely. Notice that the ace goes into the position of the 5, the 5 takes the position of the 7, which takes the position of the 8, which in turn takes the position of the ace. These four cards, 1, 5, 7, 8, are said to form a *cycle* in the permutation S. In repeating S to give the permutation 7, 3, 2, 6, 8,

4, 1, 5, which we denote by S^2, these cards follow round one another's positions cyclically, but taking two steps forward for S^2 instead of one. Thus the ace takes the position of the seven, which takes the position of the ace. Similarly, the 5 and the 8 interchange. The cycle on the four cards 1, 5, 7, 8 in S is called a *cycle of order* 4, and is denoted by (1 5 7 8). There is another cycle of order 4 in S, which is (2 4 3 6), and we write

$$S = (1\ 5\ 7\ 8)\ (2\ 4\ 3\ 6).$$

In S^2 each cycle of order 4 is replaced by two cycles of order 2, and

$$S^2 = (1\ 7)\ (5\ 8)\ (2\ 3)\ (4\ 6).$$

If m is the least common multiple of the orders of the cycles in a permutation T, then clearly T^m gives the identical permutation in which no cards are altered. This is denoted by I and $T^m = I$. m is the *order* of the permutation T.

The sum of the orders of the cycles is equal to the number of cards, if we include cycles of order one for the cards whose position is unchanged. For eight cards the greatest value of m is obtained for one cycle of order 5 and one of order 3. Thus for

$$T = (1\ 2\ 3\ 4\ 5)\ (6\ 7\ 8)$$

m is equal to 15 and the permutation has to be repeated 15 times before the original order of the cards is restored.

But even 15 is small compared with 40,320. In order to obtain a good mixing of the cards it is desirable to have a second method of shuffling that can be used in addition. For this we will take the permutation U, which takes the last card and puts it in the position of the first, to give

8, 1, 2, 3, 4, 5, 6, 7.

This gives one cycle of order 8, which is

$$U = (1\ 2\ 3\ 4\ 5\ 6\ 7\ 8).$$

If we shuffle first with the permutation S and then follow with the permutation U, the order of the cards will become

7, 8, 6, 4, 2, 1, 3, 5.

This compound permutation is denoted by SU, and in terms of the cycles

$$SU = (1\ 6\ 3\ 7)\ (2\ 5\ 8)\ 4.$$

The 4 is placed alone at the end to indicate that it is a card left unchanged, or it is sometimes omitted altogether. The rule for multiplying permutations expressed in terms of cycles is as follows:

$$\text{We have } S = (1\,5\,7\,8)\,(2\,4\,3\,6),$$
$$U = (1\,2\,3\,4\,5\,6\,7\,8).$$

To obtain the product SU notice that in S, 1 is followed by 5, and in U, 5 is followed by 6. Hence in SU, 1 will be followed by 6. In S, 6 appears at the end of the second cycle, and this is interpreted as being followed by the first element of the cycle, 2. In U, 2 is followed by 3. Hence in SU, 6 is followed by 3. Similarly $3 \to 6 \to 7$, $7 \to 8 \to 1$, and $(1\,6\,3\,7)$ completes the cycle.

If, instead of shuffling with S and then with U, we had reversed the order and shuffled first with U and then with S, we should have obtained quite a different permutation,

7, 5, 3, 1, 8, 2, 4, 6.

This compound permutation is denoted by US and in terms of cycles gives

$$US = (1\,4\,7)\,(2\,6\,8\,5).$$

Notice that SU and US are entirely different permutations. Multiplication of permutations as defined by the above rule is *non-commutative.*

These 40,320 permutations are said to form a *group.* Mathematically, a group is defined as follows:

If there is given a set of entities $a, b, c, \ldots$, with a rule for combining any two a, b, of the set to produce a third member of the set denoted by ab, then the set is said to form a *group* if the following conditions are satisfied:

(1) There exists an identity element I of the set such that $I\,a = a\,I = a$ for all elements a of the set.

(2) For each element a of the set there exists an inverse element denoted by a^{-1} such that $a\,a^{-1} = a^{-1}\,a = I$.

(3) Multiplication is associative, i.e.

$$a\,(b\,c) = (a\,b)\,c.$$

It is not assumed that multiplication is commutative.

If the number of elements of the group is finite, the group is called a *finite* group, and the number of elements is called the *order* of the group. The group we have been considering of the permutations of eight cards is clearly a group of order 40,320.

A group which permutes cards, letters, symbols, or any other set of entities is called a *permutation group*. The number of cards, letters or symbols is called the *degree* of the group. The group of all the possible permutations on r entities is of order $r(r-1)(r-2)\ldots 2.1 = r!$ and is called the *symmetric group* of degree r, or the symmetric group of order $r!$.

A group may have a subset of elements which themselves form a group. This is called a subgroup. For the symmetric group of permutations on eight cards that we have been considering, the operations S, S^2, S^3, $S^4 = I$ clearly form a group which is a subgroup of order 4. Similarly the powers of U form a subgroup of order 8. A group such as these, which consist of a single element together with its powers, is called a *cyclic* group.

If the two permutations S and U are combined, with any number of repetitions of each in any order, the resulting permutations will form a group which is either the symmetric group itself or some subgroup. It is called the group *generated* by S and U. It can be shown in this case that S and U generate the whole symmetric group, so that by a judicious mixture of the two methods of shuffling any of the 40,320 different shuffles can be obtained.

If we shuffle the pack according to the inverse of U, then with S, and finally with U to give $U^{-1} SU$ we obtain a permutation which has very similar properties to those of S. It has the same number of cycles of the same orders as for S, but the symbol in each cycle is replaced by the symbol which it replaces in the permutation U. Thus, since in U each of the numbers 1 to 7 replaces the number one greater, and 8 replaces 1, then since

$$S = (1\ 5\ 7\ 8)\ (2\ 4\ 3\ 6),$$

then the rule gives

$$U^{-1} SU = (2\ 6\ 8\ 1)\ (3\ 5\ 4\ 7).$$

The element of the group $U^{-1}SU$ is called the transform of S by the element U. For the symmetric group, but not for most other permutation groups which are subgroups of the symmetric group, any two elements which have the same number of cycles of the same order, are transforms of one another.

Thus, taking $T = (1\ 2\ 3)\ (4\ 5)\ (6\ 7)$ and $V = (8\ 6\ 2)\ (5\ 3)\ (1\ 4)$, if we choose the substitution W for which the numbers 1, 2, 3, 4, 5, 6, 7 take the places of 8, 6, 2, 5, 3, 1, 4 respectively, then $W^{-1}TW = V$ and V is the transform of T by W, while W is the transform of V by W^{-1}. This procedure might not be valid for a subgroup, because, though the required permutation W must belong to the symmetric group, it might not be a member of the given subgroup.

If V is the transform of T by any element of the group, then V and T are said to be *conjugate* elements. The set of all elements conjugate to a given element is a *class of conjugate elements*, or shortly *a class* of the group.

The classes of the symmetric groups depend on the possible arrangements of cycles, and these in turn depend upon *partitions*. If the number 8 is expressed in any way as a sum of integers, then this set of integers, irrespective of their order, is said to form a *partition* of 8. Corresponding to each partition of 8 there is a class of the symmetric group of degree 8.

Thus, since $8 = 5 + 2 + 1$, the set of numbers 5, 2, 1 forms a partition of 8 which is written shortly as $(5\ 2\ 1)$. Each element of the group which has one cycle of order 5, one cycle of order 2, with one symbol unchanged belongs to the one class corresponding to the partition $(5\ 2\ 1)$. The identity element of the group, leaving every symbol unchanged, has 8 cycles of order 1, and corresponds to the partition $8 = 1 + 1 + 1 + 1 + 1 + 1 + 1 + 1$. This partition is written as (1^8).

There are 22 partitions of 8, namely (8), (71), (62), (61^2), (53), (521), (51^3), (4^2), (431), (42^2), (421^2), (41^4), (3^22), (3^21^2), (32^21), (321^3), (31^5), (2^4), (2^31^2), (2^21^4), (21^6), (1^8). There are thus 22 classes of the symmetric group. The *order* of the class, which is the number of elements which belong to the class, can be calculated as follows.

Consider the class (32^21). Two elements of the class may be taken as $S = (123)\ (45)\ (67)\ 8$, and $T = (876)\ (54)\ (32)\ 1$. A permutation which transforms the first into the second is the one for which 1, 2, 3, 4, 5, 6, 7, 8 replaces 8, 7, 6, 5, 4, 3, 2, 1. But the cycle (876) could be written as (768) or (687).

Also the cycle (54) could be written as (45) and (32) as (23). This gives $3 \times 2 \times 2 = 12$ different ways of writing the permutation (876) (54) (32) 1, and for each of the twelve there is a permutation which transforms S into T. Further, we can interchange the two cycles (54) and (32), and this gives another twelve, so that altogether 24 elements of the group transform S into T. Similarly 24 elements transform S into any other member of the class, and since there are altogether 8 ! elements by which we can transform S, it is clear that the total number of elements in the class is $8! \div 24$.

If a is the number of cycles of order 1, b the number of cycles of order 2, c the number of order 3, etc., then the order of the class $(1^a\ 2^b\ 3^c \ldots)$ is by the same reasoning $8!/(1^a\ a!\ 2^b\ b!\ 3^c\ c! \ldots)$. The same method of reasoning shows that for any group whatsoever, the order of each class always divides exactly the order of the group.

If a group H of order h has a subgroup G of order g, then it can be shown that g always divides h exactly.

Let S_2 be an element of H which is not in G. Then if the elements of G are all multiplied by S_2 on the left, we get another set of g elements denoted by $S_2 G = G_2$, all of which are distinct from those of G. If this does not exhaust the elements of H and S_3 is a distinct element, then $S_3 G = G_3$ is a third set of g elements, all of which are distinct from those of G and G_2. Proceeding in the same way until all elements of H are exhausted, we obtain say ν sets each of g elements, $G, G_2, G_3, \ldots, G_\nu$, which are called *cosets* of G, and $g\ \nu = h$, so g divides h exactly.

If the cosets

$$G, S_1 G, S_2 G, \ldots, S_\nu G$$

are multiplied on the left by any element T of the group, then each coset is changed into another (or the same) coset. If TS_i is a member of $S_j G$, then TG_i becomes G_j. In this way, multiplying by T on the left effects a permutation of the cosets. Corresponding to another element U, multiplying on the left effects another permutation of the cosets, and the product of these permutations corresponds to the product UT of the elements of the group.

Each subgroup G of H, therefore, leads to a representation of H as a permutation group of degree $\nu = h/g$. The importance of this result follows from the fact that almost every group possesses a subgroup; most of them have numerous subgroups. The only groups without subgroups are cyclic groups of prime order. Even in this case we can consider that the identity element by itself forms a subgroup of order one. It certainly does define in the same way a permutation representation whose degree is equal to the order of the group, and which is called the *regular permutation representation of the group.*

Hence a knowledge of the permutation groups implies a knowledge of all possible finite groups, since they all have representations as permutation groups.

CHAPTER IX

THE GALOIS THEORY OF EQUATIONS

SUPPOSE that there is given a cubic equation, say
$$x^3 + a x^2 + b x + c = 0.$$
This equation will always have three roots, which will be real or complex numbers, and which we will denote by α, β, γ.

To find α, β, γ it is of course necessary to solve the equation, and this may not be an easy procedure, but there are certain functions of the roots α, β, γ which can be determined easily without solving the equation.

If we are given a function of the roots, say $\alpha^2 \beta$, we can in general form five other functions, e.g. $\alpha^2 \gamma$, $\beta^2 \alpha$, $\beta^2 \gamma$, $\gamma^2 \alpha$, $\gamma^2 \beta$ by taking a different arrangement of the roots α, β, γ, i.e. by operating on the function with a permutation interchanging the roots. The six operations of the symmetric group give six functions which are all distinct in this case.

It may happen that all the operations of the group leave the function unchanged, as, e.g. when the function is $(\alpha + \beta + \gamma)$ or $\alpha \beta \gamma$. Such functions are called symmetric functions. Any symmetric function of the roots can be expressed directly in terms of the coefficients in the equation without solving the equation.

Thus the equation
$$x^3 + a x^2 + b x + c = 0$$
must be exactly the same as the equation
$$(x - \alpha)(x - \beta)(x - \gamma) = 0,$$
and expanding the latter and comparing the coefficients of the various powers of x,
$$\alpha + \beta + \gamma = -a,$$
$$\alpha \beta + \beta \gamma + \gamma \alpha = b,$$
$$\alpha \beta \gamma = -c.$$

In terms of these any other symmetric function can be expressed. Thus

$$\alpha^2 + \beta^2 + \gamma^2 = (\alpha + \beta + \gamma)^2 - 2(\alpha\beta + \beta\gamma + \gamma\alpha)$$
$$= a^2 - 2b.$$

Now it may happen that we can find a function of the roots that is changed by some of the permutations but is left unchanged by others. Then the permutations which leave it unchanged will form a subgroup of the symmetric group, and the function is said to *belong to this subgroup.*

Thus there is a subgroup of order two which contains the identity and the interchange $(\alpha\beta)$. Examples of functions which belong to this subgroup are $(\alpha^2 + \beta^2)$, $(\alpha + \beta)$, γ, $\alpha\gamma + \beta\gamma$.

These functions of the roots cannot be expressed in terms of the coefficients a, b, c, without solving the equation, but they have the remarkable property that any one of them can be expressed in terms of any other, with the aid of the coefficients.

Thus
$$(\alpha + \beta) = -a - \gamma,$$
$$(\alpha^2 + \beta^2) = (\alpha + \beta)^2 - 2\alpha\beta$$
$$= (a + \gamma)^2 - 2c/\gamma,$$
$$\alpha\gamma + \beta\gamma = -a\gamma - \gamma^2.$$

To express, say, γ in terms of $(\alpha^2 + \beta^2)$ is not so easy, but can nevertheless be accomplished. Thus

$$\gamma^2 = \alpha^2 + \beta^2 + \gamma^2 - (\alpha^2 + \beta^2) = a^2 - 2b - (\alpha^2 + \beta^2),$$

and since $\gamma^3 + a\gamma^2 + b\gamma + c = 0,$

then $\gamma(\gamma^2 + b) = -a\gamma^2 - c,$

so that $\gamma = -(a\gamma^2 + c)/(\gamma^2 + b).$

This expresses γ in terms of γ^2, which in its turn has already been expressed in terms of $(\alpha^2 + \beta^2)$.

Now there are three subgroups of the symmetric group, each of order two, and of the same type as the one we are considering. These are:

$$G_1;\ I,\ (\alpha\beta);$$
$$G_2;\ I,\ (\beta\gamma);$$
$$G_3;\ I,\ (\gamma\alpha).$$

Any of these subgroups can be transformed into any other by an operation of the symmetric group. They are called *conjugate subgroups*, and the set of three is a *class of conjugate subgroups.*

The other subgroup of the symmetric group of order six is the subgroup of order three consisting of the elements

$$G;\ I,\ (\alpha\,\beta\,\gamma),\ (\alpha\,\gamma\,\beta).$$

This is different because there is no other subgroup conjugate to it. Every transform of G gives the same subgroup G, and it is called a *self conjugate subgroup.*

An example of a function of the roots which belongs to this subgroup is

$$(\alpha-\beta)(\alpha-\gamma)(\beta-\gamma)=\alpha^2\beta+\beta^2\gamma+\gamma^2\alpha-\alpha\beta^2-\beta\gamma^2-\gamma\alpha^2.$$

The ratio of the order of the group to the order of the subgroup is called the *index* of the subgroup.

We will now consider how the properties of these subgroups can be used in the solution of equations.

It is well known that, in general, the solution of an algebraic equation of degree greater than one involves irrational numbers. Hence to find the solution, even of a quadratic, some process must be employed which obtains an irrational number. The simplest process which yields an irrational number is the extraction of roots, that is the finding of the square root, cube root or n-th root of a given number. We say that an equation is *solvable* if by finding n-th roots of numbers a certain number of times we can reach an expression which satisfies the equation. It is well known that any quadratic equation can be solved by the extraction of one square root. Hence the quadratic equation is solvable.

Now if those functions of the roots of an equation which correspond to a given group H are known, and the group H has a subgroup G of index r, then the functions which belong to the group G can be obtained by solving an equation of degree r. Further, if G is a self conjugate subgroup of H and r is a prime number, it can be shown that this equation can be put in the form

$$x^r = k,$$

and thus the step from the functions belonging to H to the functions belonging to G can be made by the extraction of one r-th root. Hence, if H is the symmetric group on n

symbols, we can solve the n-th degree equation if we can find a sequence of subgroups

$$H, G_1, G_2, \ldots, G_r = I$$

ending in the group of order one which consists of the identity, such that each group is an invariant subgroup of the preceding group, of prime index. Further, only when such a sequence of subgroups can be obtained is the general n-th degree equation solvable.

For the cubic equation such a sequence exists, for we can take H as the symmetric group, then G as the invariant subgroup of order 3, and finally the identity.

A function belonging to the group G is

$$D = (\alpha - \beta)(\alpha - \gamma)(\beta - \gamma) = \alpha^2\beta + \beta^2\gamma + \gamma^2\alpha - \alpha\beta^2 - \beta\gamma^2 - \gamma\alpha^2.$$

Clearly D^2 is a symmetric function. This can be expressed in terms of a, b, c and is in fact

$$D^2 = -27c^2 - 4b^3 + a^2b^2 + 18abc - 4a^3c.$$

From this D is found by the extraction of the root, and since $\alpha^2\beta + \beta^2\gamma + \gamma^2\alpha + \alpha\beta^2 + \beta\gamma^2 + \gamma\alpha^2$ is a symmetric function and hence is known, we can also find $\alpha^2\beta + \beta^2\gamma + \gamma^2\alpha$, and all such functions as are unchanged when α, β, γ are permuted cyclically.

It is now required to find, say, $\alpha^2\beta$ by the extraction of a cube root. The permutation S which permutes α, β, γ cyclically, satisfies $S^3 = I$. We therefore make it correspond to the complex number ω, which is a cube root of unity, $\omega = \frac{1}{2}(-1 + i\sqrt{3})$.

We obtain thus from $\alpha^2\beta$ by the operator $(I + \omega S + \omega^2 S^2)$ the quantity

$$U = \alpha^2\beta + \omega\beta^2\gamma + \omega^2\gamma^2\alpha.$$

Then

$$\begin{aligned} U^3 = {} & \alpha^6\beta^3 + \beta^6\gamma^3 + \gamma^6\alpha^3 \\ & + 3\omega\alpha\beta\gamma(\alpha^3\beta^2 + \beta^3\gamma^2 + \gamma^3\alpha^2) \\ & 3\omega^2\alpha\beta\gamma(\alpha^2\beta^3 + \beta^2\gamma^3 + \gamma^2\alpha^3) \\ & 6\alpha^3\beta^3\gamma^3. \end{aligned}$$

These expressions all belong to the group G and can all be evaluated. Further, since $\alpha^2\beta$ belongs to the group of order one, each of α, β, γ can be expressed in terms of it.

Now this method would be a very clumsy and laborious

method to use in practice. Neat and concise methods like the following are generally employed. For the general cubic

$$x^3 + a x^2 + b x + c = 0$$

replace $(x + \frac{1}{3} a)$ by x to obtain a cubic in which the term in x^2 is absent, say

$$x^3 + p x + q = 0.$$

Then if $x = y + z$ the equation can be written

$$y^3 + z^3 + 3 y z (y + z) + p (y + z) + q = 0.$$

Further, suppose that y and z are restricted so that

$$3 y z + p = 0.$$

Then

$$y^3 + z^3 = -q,$$

$$y^3 z^3 = -p^3/27,$$

so that y^3 and z^3 are the roots of

$$\lambda^2 + q \lambda - p^3/27 = 0$$

regarded as a quadratic equation in λ. When y^3 is found from this by the extraction of a square root, then y can be found from y^3 by the extraction of a cube root. Further $z = -p/3y$, and the three roots of the cubic are

$$y + z,\ \omega y + \omega^2 z,\ \omega^2 y + \omega z$$

where ω is the complex cube root of unity.

These two methods are, however, fundamentally the same. The second is just a simplified and neat form of the first. It applies only to the cubic, however, and has no application to any other equation.

The first method, however, is perfectly general and can be used to solve any solvable equation.

For the quartic equation with roots α, β, γ, δ, the appropriate group is the symmetric group on the four symbols α, β, γ, δ. The 24 operations of this group separate into five classes which correspond to the partitions of 4. The orders of the classes are as follows:

Class: (1^4), $(1^2 2)$, (13), (4), (2^2).
Order: 1, 6, 8, 6, 3.

This symmetric group has a self-conjugate subgroup of order 12, which consists of the classes (1^4), (13) and (2^2). We denote the symmetric group by H, and this group of order 12, which is called the *alternating group*, by G_1.

The group G_1 has a self-conjugate subgroup of order 4, which we denote by G_2, which is composed of the elements from the classes (1^4) and (2^2), i.e. the elements I, $(\alpha\,\beta)\,(\gamma\,\delta)$, $(\alpha\,\gamma)\,(\beta\,\delta)$, $(\alpha\,\delta)\,(\beta\,\gamma)$. These elements are all self-conjugate in G_2, and hence there are self-conjugate subgroups of order 2, e.g. G_3 consisting of I and $(\alpha\,\beta)\,(\gamma\,\delta)$.

To solve a quartic equation with roots α, β, γ, δ we obtain first a function of the roots belonging to G_1. The simplest of such functions is

$$\begin{aligned} D = {} & \alpha^3\beta^2\gamma + \alpha^3\gamma^2\delta + \alpha^3\delta^2\beta + \beta^3\gamma^2\alpha + \beta^3\alpha^2\delta + \beta^3\delta^2\gamma \\ & + \gamma^3\alpha^2\beta + \gamma^3\beta^2\delta + \gamma^3\delta^2\alpha + \delta^3\alpha^2\gamma + \delta^3\gamma^2\beta + \delta^3\beta^2\alpha \\ & - \alpha^3\gamma^2\beta - \alpha^3\delta^2\gamma - \alpha^3\beta^2\gamma - \beta^3\alpha^2\gamma - \beta^3\delta^2\alpha - \beta^3\gamma^2\delta \\ & - \gamma^3\beta^2\alpha - \gamma^3\delta^2\beta - \gamma^3\alpha^2\delta - \delta^3\gamma^2\alpha - \delta^3\beta^2\gamma - \delta^3\alpha^2\beta \\ = {} & (\alpha-\beta)\;(\alpha-\gamma)\;(\alpha-\delta)\;(\beta-\gamma)\;(\beta-\delta)\;(\gamma-\delta). \end{aligned}$$

Then D^2 is clearly a symmetric function and can be expressed in terms of the coefficients in the equation, and extraction of the square root gives the value of D.

Putting $D = D_1 - D_2$ where these represent respectively the positive and negative terms, clearly $D_1 + D_2$ is a symmetric function expressible in terms of the known coefficients and hence can be found D_1 and D_2.

The simplest function* belonging to the group G_2 is $(\alpha\,\beta + \gamma\,\delta)$. This group is of index 3 in G_1 and the conjugate expressions are $(\alpha\,\gamma + \beta\,\delta)$, $(\alpha\,\delta + \beta\,\gamma)$. Hence put

$$Z = (\alpha\,\beta + \gamma\,\delta) + \omega\,(\alpha\,\gamma + \beta\,\delta) + \omega^2\,(\alpha\,\delta + \beta\,\gamma)$$

where ω is the complex cube root of unity $\frac{1}{2}(-1 + i\sqrt{3})$. Then

$$\begin{aligned} Z^3 = {} & \Sigma\,\alpha^3\beta^3 + 3\,\Sigma\,\alpha^2\beta^2\gamma\,\delta + \omega\,[D_1 + 2\,\alpha\,\beta\,\gamma\,\delta\,\Sigma\,\alpha\,\beta] \\ & \omega^2\,[D_2 + 2\,\alpha\,\beta\,\gamma\,\delta\,\Sigma\,\alpha\,\beta]. \end{aligned}$$

This can be expressed in terms of the symmetric functions and D. Hence Z can be found by the extraction of a cube root. From Z the values of $(\alpha\,\beta + \gamma\,\delta)$, $(\alpha\,\gamma + \beta\,\delta)$, $(\alpha\,\delta + \beta\,\gamma)$ can be deduced.

To proceed to the group G_3 we put

$$w = \alpha\,\beta - \gamma\,\delta.$$

Then $w^2 = (\alpha\,\beta + \gamma\,\delta)^2 - 4\,\alpha\,\beta\,\gamma\,\delta$ is known, and the extraction of a square root gives the value of w, from which $\alpha\,\beta$ may be determined. The function $(\alpha + \beta)$ belonging to the

* Strictly, this function belongs to a group of order 8 which is mentioned below, and which includes G_2 as a subgroup. The other 4 operations are excluded, however, by the use of D.

same subgroup as $\alpha\beta$ can be expressed in terms of $\alpha\beta$, and the roots α and β are determined by solving the known quadratic equation

$$x^2 - (\alpha + \beta)\, x + \alpha\beta = 0.$$

This shows that any quartic equation can be solved by the extraction of roots. Once again, to use the above method as described to solve a given numerical quartic equation would prove awkward and cumbersome, but once it is known that an equation is solvable it is comparatively straightforward to devise a neater and more usable method.

In actual practice it is more convenient to proceed from the symmetric group to the group of order 8 comprising

$$I,\ (\alpha\beta),\ (\gamma\delta),\ (\alpha\beta)\ (\gamma\delta),\ (\alpha\gamma)\ (\beta\delta)$$
$$(\alpha\delta)\ (\beta\gamma),\ (\alpha\gamma\beta\delta),\ (\alpha\delta\beta\gamma).$$

This group is of index 3 in the symmetric group, but is not self-conjugate, and gives rise to a cubic equation called the *auxiliary cubic.* It is not of the form $y^3 = c$, since the subgroup is not self-conjugate, but is quite a general cubic equation. It is solvable since every cubic is solvable.

The group of order 8 has the self-conjugate subgroup consisting of I, $(\alpha\beta)$, $(\gamma\delta)$, $(\alpha\beta)\ (\gamma\delta)$. The transition to this subgroup is equivalent to the factorization of the quartic into two quadratics, from which the solution follows.

Thus if the quartic is

$$x^4 + 4\,a\,x^3 + 6\,b\,x^2 + 4\,d\,x + e = 0,$$

this is put in the form

$$(x^2+2\,a\,x+y)^2-[(4a^2+2g-6b)\,x^2+(4ay-4d)\,x+y^2-e]=0.$$

If y satisfies a certain cubic equation, the auxiliary cubic, then the expression in the square bracket becomes an exact square, so that the equation is of the form

$$(x^2 + 2ax + y)^2 - (px + q)^2 = 0,$$

whence the factorization and solution follows.

Notice that $y = \frac{1}{2}\,(\alpha\beta + \gamma\delta)$, which is a function of the roots belonging to the above group of order 8.

The appropriate group for the general quintic or fifth degree equation is the symmetric group of order $5! = 120$. This group has a self-conjugate subgroup of order 60, the alternating group. But it can be shown that this alternating group has no invariant subgroup. Hence the procedure fails, and the general quintic equation it is impossible to solve by the repeated extraction of roots.

Special quintic equations, however, it may be possible to solve. The symmetric group has a subgroup of order 20 which is generated by the cyclic interchange $(\alpha\,\beta\,\gamma\,\delta\,\epsilon)$ on the five roots and the cycle $(\beta\,\gamma\,\epsilon\,\delta)$ on four. A function of the roots belonging to this subgroup is

$$z = \alpha^2(\beta\gamma + \gamma\epsilon + \delta\beta) + \beta^2(\gamma\delta + \delta\alpha + \epsilon\gamma) + \gamma^2(\delta\epsilon + \epsilon\beta + \alpha\delta) + \delta^2(\epsilon\alpha + \alpha\gamma + \beta\epsilon) + \epsilon^2(\alpha\beta + \beta\delta + \gamma\alpha).$$

Since the subgroup is of index 6 in the symmetric group, z must satisfy a certain equation of degree 6. Such an equation is called a Galois resolvant. If it should happen that this equation has a rational root, then the equation is said to belong to this group of order 20. It can then be shown that the equation is solvable by a process of repeated extraction of roots, for an examination of this group of order 20 shows that it is a solvable group.

For the general equation of degree n, the appropriate group is the symmetric group on n symbols. Corresponding to each subgroup there is a Galois resolvant equation. If this equation has a rational root then the equation is said to belong to this subgroup. If the group is a solvable group, then the given equation can be solved by a repeated extraction of roots. Under no other circumstances can the equation be solved in this way.

For a given subgroup the Galois resolvant equation is not unique, for a different function of the roots belonging to this subgroup could be found, and this would satisfy a different equation. But if one resolvant corresponding to the subgroup has a rational root, then every resolvant corresponding to the subgroup will have a rational root. From the point of view of solving the equation, all such resolvant equations may be regarded as equivalent.

Among the class of equations whose groups are less than the symmetric group a notable example is furnished by the reciprocal equation. This is an equation whose coefficients are the same when read in reverse order. Thus the reciprocal sixth degree equation is of the form

$$x^6 + a\,x^5 + b\,x^4 + c\,x^3 + b\,x^2 + a\,x + 1 = 0.$$

Clearly the same equation is obtained if x is replaced by $1/x$. Hence if α is a root, so also is $1/\alpha$. Instead of six quite independent roots there are three pairs of roots:

$$\alpha, \quad \alpha', \qquad \beta, \quad \beta', \qquad \gamma, \quad \gamma'$$

where $\alpha' = 1/\alpha, \quad \beta' = 1/\beta, \quad \gamma' = 1/\gamma.$

The group of such an equation allows for a possible interchange of each pair of roots, e.g. α with α', or for any permutation of the three sets of pairs. It has thus a subgroup of order 8 which allows of an interchange of each pair without permutating the pairs. This invariant subgroup is of index 6, which is the same as the degree of the symmetric group on three symbols, and the relation of the invariant subgroup to the full group is exactly the same as the relation of the identical element to the symmetric group on three symbols. We say that the *quotient group* of the group by the self-conjugate subgroup is the symmetric group on three symbols. The resolvent of the reciprocal equation corresponding to the invariant subgroup of order 8 is an ordinary cubic equation. The solution of the group of order 8 is clearly equivalent to the solution of three separate quadratics. Hence the reciprocal equation of degree 6 can be reduced to three separate quadratics by the solution of one cubic equation.

The form of the solution is as follows. Divide the equation by x^3 to give

$$x^3 + \frac{1}{x^3} + a\left(x^2 + \frac{1}{x^2}\right) + b\left(x + \frac{1}{x}\right) + c = 0.$$

Then put $\qquad y = x + \dfrac{1}{x}$

to give $\qquad y^2 = x^2 + \dfrac{1}{x^2} + 2,$

$$y^3 = x^3 + \frac{1}{x^3} + 3\,x + \frac{3}{x}.$$

Thus

$$y^3 - 3y + a\,(y^2 - 2) + b\,y + c = 0.$$

The solution of this cubic gives the value of $x + \dfrac{1}{x}$, which belongs to the group of order 8, and the equation reduces to three separate quadratics.

A problem which arises in geometry concerns the construction of certain angles by the use of a rule and compasses. It is well known that such angles as 60°, 90°, 45° can be obtained by rule and compass constructions, and Euclid gives also a construction involving what is called sometimes "golden section" which leads to the angles of 18° and 36°. A complete account of all those submultiples of 360° which can be constructed by rule and compasses can be deduced from the Galois theory.

Owing to the fact that Pythagoras' theorem involves the areas of the *squares* on the sides, constructions with rule and compasses can be used to construct geometrically any square root. Further, the equivalent of this procedure is the only method of constructing irrational lengths with rule and compasses.

Hence, if the unit length is given, a length x can be constructed by rule and compasses if, and only if, x is the root of an equation which is not only solvable, but is also solvable by a sequence of quadratic equations. This is possible if the group to which x belongs has a sequence of subgroups ending in the identity such that each is an invariant subgroup of index 2 in the preceding group. For this to happen the order of the group must be a power of 2.

The angle $360°/n$ is that associated in the Argand diagram with a complex root of the equation $x^n = 1$. Thus the angle $360°/n$ can be constructed with rule and compasses if, and only if, this root x belongs to a solvable group whose order is a power of 2.

Firstly if $n = 2^r$ this condition is satisfied. Such an angle $360°/2^r$ can be easily constructed by repeatedly bisecting a right angle.

Secondly, if $n = p^2$ where p is an odd prime, the group to which a primitive root of $x^{p^2} = 1$ belongs can be shown to be divisible by p, and hence no angle of magnitude $360°/p^2$ can be constructed with rule and compasses.

Otherwise, since when p and q are co-prime, the fraction

$1/pq$ can be put in the form $\frac{\alpha}{p} + \frac{\beta}{q}$, it is sufficient to consider the case when $n = p$, an odd prime. Removing the obvious factor $x - 1$ from the equation $x^p - 1 = 0$, the degree of the equation obtained is $p - 1$. This equation can be solved by a sequence of quadratics if, and only if, $p - 1$ is a power of 2.

If $2^r + 1$ is a prime number, then the equation $x^{2^r+1} - 1 = 0$ can be solved by a sequence of quadratics, and the angle $360°/(2^r + 1)$ can be constructed by rule and compasses.

The simplest examples of prime numbers of this form are 3, 5, 17, 257, and the following angles can be so constructed.

The construction of the angle $360°/3 = 120°$ is very well known. The angle $360°/5 = 72°$ is dealt with by Euclid's "golden section". The angles $360°/17$ and $360°/257$ can also be constructed in a similar manner.

The manner of solution of the equation $x^{17} - 1 = 0$ will now be briefly indicated. If α is a complex root, then $\alpha, \alpha^2, \alpha^3, \ldots, \alpha^{16}, \alpha^{17} = 1$ are the roots of the equation. We place the complex roots in a sequence beginning with α by putting after each root its cube, thus:

$$\alpha, \alpha^3, \alpha^9, \alpha^{10}, \alpha^{13}, \alpha^5, \alpha^{15}, \alpha^{11}, \alpha^{16}, \alpha^{14}, \alpha^8, \alpha^7, \alpha^4, \alpha^{12}, \alpha^2, \alpha^6.$$

The sum of these roots is -1. Putting z equal to the sum of the alternate roots

$$z = \alpha + \alpha^9 + \alpha^{13} + \alpha^{15} + \alpha^{16} + \alpha^8 + \alpha^4 + \alpha^2,$$

then $-1 - z = \alpha^3 + \alpha^{10} + \alpha^5 + \alpha^{11} + \alpha^{14} + \alpha^7 + \alpha^{12} + \alpha^6$.

Subtracting and squaring $(1 + 2z)^2$ can be shown to be a symmetric function, and hence rational. It is in fact equal to 17. Hence $z = \frac{1}{2}(\sqrt{17} - 1)$.

Taking the alternate roots in z, put

$$w = \alpha + \alpha^{13} + \alpha^{16} + \alpha^4,$$
$$z - w = \alpha^9 + \alpha^{15} + \alpha^8 + \alpha^2.$$

Subtracting and squaring, it can be shown that

$$(2w - z)^2 = 8 - z,$$

whence $$w = \tfrac{1}{2}[z + \sqrt{(8 - z)}].$$

Proceeding similarly

$$u = \alpha + \alpha^{16}, \quad w - u = \alpha^{13} + \alpha^4,$$
$$(2u - w)^2 = -2w^2 - 3w + 3z + 8.$$

When u is known a is the root of

$$a^2 - u a + 1 = 0.$$

To translate this result into geometrical terms it is just necessary to show how to solve the quadratic equation $x^2 - 2ax + b = 0$ by geometrical construction. Draw a line ABC such that AB is b units and BC one unit. Let

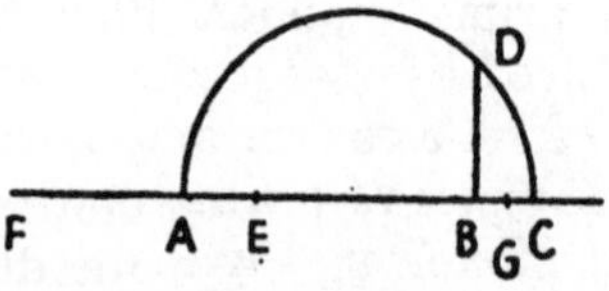

the perpendicular at B meet the circle on AC as diameter in D. With compasses, mark a point E on AC such that DE is a units and mark points F and G on AC produced if necessary such that $FE = EG = ED = a$ units. Then the lengths FB and BG give the roots of the equation, as may be readily verified by the use of Pythagoras' theorem.

CHAPTER X

ALGEBRAIC GEOMETRY

A METHOD of studying Geometry by the use of Algebra was discovered by Descartes. Considering simply the Geometry of a plane, the method may be described as follows.

Take any fixed point O in the plane, and two other fixed points AB not collinear with O. The motion which moves all points of the plane a distance OA in the direction parallel to OA is called a displacement and is denoted by OA. The

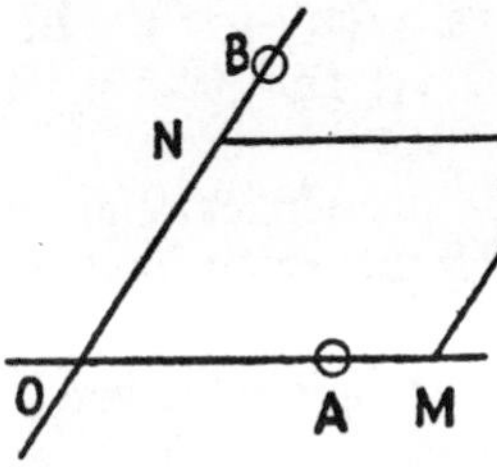

resultant of two such displacements is a third displacement which is called the *sum* of the displacements, and the addition sign $+$ is used for such a resultant. A displacement parallel to OA but of magnitude a times OA is denoted by $a\,OA$. The numerical coefficient is taken as negative if the direction is reversed.

Then if P is any point in the plane, the displacement OP can be expressed as a displacement parallel to OA together with a displacement parallel to OB, thus

$$OP = OM + MP.$$

Then numbers x, y can be found such that $OM = x\,OA$ and $MP = y\,OB$. Thus

$$OP = x\,OA + y\,OB.$$

The pair of numbers (x, y) are uniquely determined by the point P, and conversely the pair of numbers uniquely determine the point P. They are called the ***co-ordinates*** of the point P.

The commutative relation $x\,OA + y\,OB = y\,OB + x\,OA$ is equivalent to the theorem of Euclid, that the opposite sides of a parallelogram are equal and parallel. It is often convenient

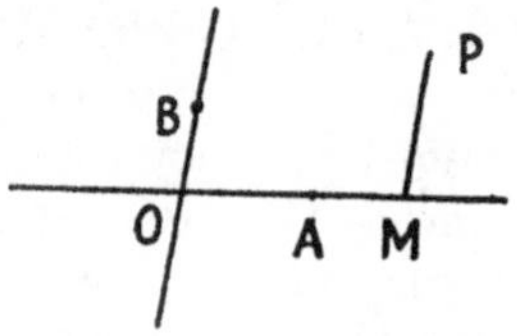

to take OA and OB to be of the same length and perpendicular to one another. If OA is then taken as the unit of length, by Pythagoras' theorem

$$OP^2 = OM^2 + MP^2$$

and the length of OP is $\sqrt{(x^2 + y^2)}$. Similarly the distance between the points (x_1, y_1) and (x_2, y_2) is $\sqrt{[(x_1 - x_2)^2 + (y_1 - y_2)^2]}$.

Various geometrical properties can thus be expressed algebraically in terms of the co-ordinates of the constituent points.

Thus the area of the triangle with vertices (x_1, y_1), (x_2, y_2) and (x_3, y_3) can be expressed as

$$\tfrac{1}{2}(x_1y_2 - x_2y_1 + x_2y_3 - x_3y_2 + x_3y_1 - x_1y_3).$$

The set of points which lie on a given line all satisfy an equation of the first degree such as $a\,x + b\,y + c = 0$, which is called the equation to the line. If this equation is put in the form $y = m\,x + c$, then m represents the *gradient* of the line, represented trigonometrically by tan θ where θ is the angle the line makes with the line OA which is called the x-axis.

If two lines have the same gradient they are parallel. If two lines have gradients m and m' respectively they will be perpendicular if $m\,m' = -1$.

An equation of degree greater than one represents, in general, a curve. If the degree is two, the curves are circles or those curves studied by Euclid as the plane sections of a circular cone, namely the ellipse, hyperbola and parabola. Thus for a circle with centre origin, a point on the curve is at distance r from the origin and thus

$$x^2 + y^2 = r^2.$$

The general circle centre (α, β) and radius r is

$$(x - \alpha)^2 + (y - \beta)^2 = r^2.$$

By comparison the equation of the form

$$x^2 + y^2 + 2gx + 2fy + c = 0$$

represents a circle centre $(-g, -f)$ and radius $\sqrt{(f^2 + g^2 - c)}$.

If (x_1, y_1) represents a point on the curve, then it can be shown that the tangent line at this point is obtained from the equation to the circle by replacing the term x^2 by $x x_1$ and similarly y^2 by $y y_1$ and also each term $2x$ by $x + x_1$ and $2y$ by $y + y_1$. Thus

$$x x_1 + y y_1 + g(x + x_1) + f(y + y_1) + c = 0.$$

But if (x_1, y_1) is any point, not necessarily on the circle, then this equation still represents a line, not a tangent line, but a line which has a certain geometrical relation to the point and the circle. It is called the *polar* of the point.

Although the polar is obtained with reference to the particular system of co-ordinates used, it is in a manner independent of the co-ordinate system. If an entirely different system of co-ordinates is used, suppose that the co-ordinates of a point are (x', y'), of the fixed point (x_1', y_1'), and that the equation to the circle is

$$x'^2 + y'^2 + 2g' x' + 2f' y' + c' = 0.$$

Then it can be shown that the polar line

$$x' x_1' + y' y_1' + g'(x' + x_1') + f'(y' + y_1') + c' = 0$$

represents exactly the same line as the polar line $x x_1 + y y_1 + g(x + x_1) + f(y + y_1) + c = 0$ referred to the original co-ordinates. Such an operation as finding the polar of a point with respect to a circle is called an *invariant* operation. The polar is a *concomitant* of the point and the circle. Since the result is independent of the co-ordinate system it must indicate some geometrical relation. It is easy to find in the case of pole and polar, the nature of this relation.

The condition that the polar of (x_1, y_1) passes through (x_2, y_2) is that

$$x_1 x_2 + y_1 y_2 + g(x_1 + x_2) + f(y_1 + y_2) + c = 0.$$

From the symmetry of this relation it follows that if the polar of a point P passes through a point Q, then the polar of Q

passes through P. Also the polar of a point on the circle is the tangent at this point.

If P is a point outside the circle, draw the tangents PT, PT' to the circle. Then since the polar of T passes through P, the polar of P will pass through T, and similarly T'. Hence the polar of P is the line TT'.

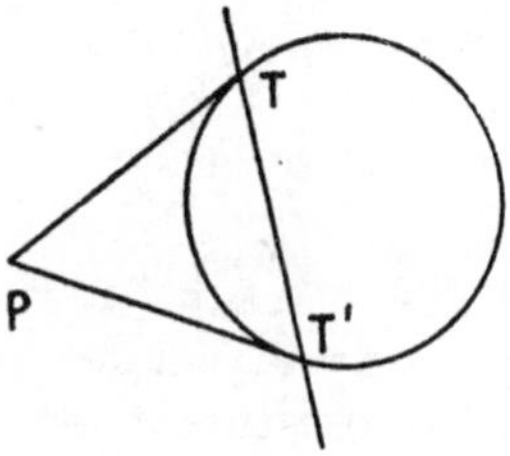

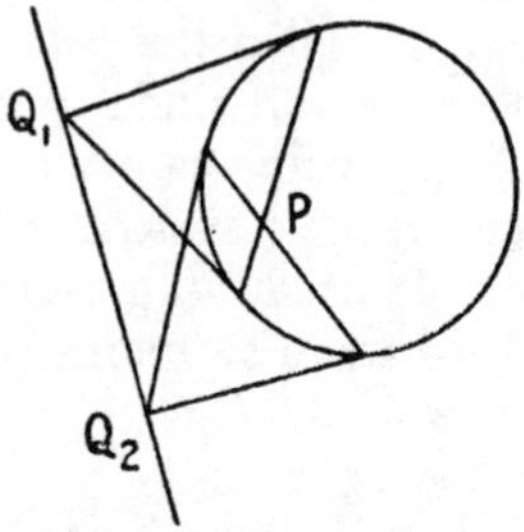

For a point P in the circle, draw any chord through P and let the tangents at the extremities meet in Q_1. If the corresponding tangents for another chord meet in Q_2, then the line $Q_1 Q_2$ is the polar of P.

The polar has other geometrical properties which can be demonstrated by the algebraic manipulation of the co-ordinates. Thus if any line through P meets the polar of P in Q, then the circle cuts PQ in two points which divide it internally and externally in the same ratio.

The other second degree curves, the conic sections, or shortly, conics, include the ellipse, of which the equation in its simplest form is

$$\frac{x^2}{a^2} + \frac{y^2}{b^2} = 1,$$

the hyperbola, with equation

$$\frac{x^2}{a^2} - \frac{y^2}{b^2} = 1,$$

and the parabola $y^2 = 4ax$.

The general second degree equation can be written

$$ax^2 + by^2 + 2hxy + 2gx + 2fy + c = 0.$$

The equation to the tangent at a point (x_1, y_1) on the curve can be obtained from this by replacing x^2 by $x x_1$, and $2x$ by $(x + x_1)$ as for the circle, while the extra term $2xy$ is replaced by $(x y_1 + x_1 y)$.

The polar of a point is defined exactly as for circles, and has the same properties.

By a judicious change in the co-ordinate system the equation can be brought either to the form

$$y^2 = 4a' x,$$

which indicates that the curve is a parabola, or else to the form

$$a' x^2 + b' y^2 + c' = 0.$$

If $c' = 0$ this represents the pair of lines

$$\sqrt{a'}\, x = \pm \sqrt{-b'}\, y,$$

which may be real or imaginary, according as a and b are of the opposite or of the same sign. If c' is not zero then either there is no real curve if a', b', c' have the same sign, or else if a', b' have the same sign different to that if c' it is an ellipse, or if a', b' have opposite signs, a hyperbola.

In order to find the nature of the curve from the general equation one can use the properties of the invariants.

Firstly it can be shown that the expression

$$\Delta = abc - af^2 - bg^2 - ch^2 + 2fgh$$

remains invariant for any change of the co-ordinates. Hence it is called an *invariant*. Being independent of the co-ordinate system, the equation $\Delta = 0$ represents some geometrical property of the curve. It is in fact the condition that the equation represents a pair of straight lines.

Secondly it can be shown that the equation

$$\lambda^2 - (a + b)\lambda + ab - h^2 = 0$$

is an invariant equation, so that the coefficients $(a + b)$ and $ab - h^2$ are invariants.

Setting aside for the moment the case of the parabola, for the simpler form of the curve

$$a'\,x^2 + b'\,y^2 + c' = 0,$$

which is called the *canonical form*, the invariant equation becomes

$$\lambda^2 - (a' + b')\,\lambda + a'\,b' = 0.$$

Hence the values of a' and b' can be found by obtaining the roots of the equation

$$\lambda^2 - (a + b)\,\lambda + a\,b - h^2 = 0.$$

Then, since

$$c' = a'\,b'\,c'/a'\,b' = \Delta/(a\,b - h^2),$$

the value of c' can be found from the invariants, and the canonical form of the curve and its exact nature can be deduced.

The case of the parabola arises when $a\,b - h^2 = 0$. Then the canonical form of the curve is taken as

$$a'\,x^2 + 2f'\,y = 0,$$

and the values of a' and f' can be deduced from the invariants

$$a' = a + b,$$
$$-\,a'\,f'^2 = \Delta.$$

Generally speaking, the algebraic forms obtained in algebraic geometry represent relations between the geometric configurations and the system of co-ordinates. The use of invariants and of other concomitants is equivalent to the elimination of the co-ordinate system, and concomitant forms represent pure geometric properties.

Many geometrical results which can be demonstrated neatly and elegantly by pure geometrical methods can alternatively be proved by the methods of algebraical geometry by considering the co-ordinates of points and by performing heavy and laborious calculation with them. Consequently algebraic geometry has come to be regarded by many as the crude stand-by of an inferior mathematician who can perform laborious calculation, but is unable to master the elegant methods of pure geometry.

Such a view, however, takes no account of the higher refinements of algebraic geometry. By a judicious use of invariants and other concomitants, and of matrices, which will be dealt with in the next chapter, methods and proofs can be obtained which are as concise and elegant as any results of pure geometry, and which are more far-reaching in their scope.

Before concluding this chapter, some account of the method of homogeneous co-ordinates would not be out of place.

Let *ABC* be three fixed points in a plane with co-ordinates (x_1, y_1), (x_2, y_2), (x_3, y_3). The co-ordinates of the point *L* which divides *BC* in the ratio $\nu : \mu$ has co-ordinates

$$\left(\frac{\mu x_2 + \nu x_3}{\mu + \nu}, \frac{\mu y_2 + \nu y_3}{\mu + \nu}\right).$$

The ratio $\nu : \mu$ is taken as negative if *BC* is divided externally instead of internally.

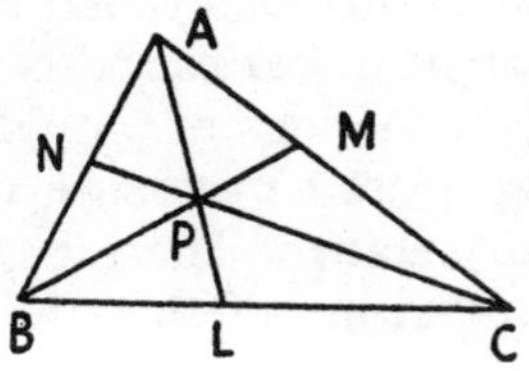

The point *P* which divides *AL* in the ratio $(\mu + \nu) : \lambda$ has co-ordinates

$$\left(\frac{\lambda x_1 + \mu x_2 + \nu x_3}{\lambda + \mu + \nu}, \frac{\lambda y_1 + \mu y_2 + \nu y_3}{\lambda + \mu + \nu}\right).$$

Then clearly if *B* meets *AC* in *M*, then *M* is the point

$$\left(\frac{\nu x_1 + \lambda x_3}{\nu + \lambda}, \frac{\nu y_1 + \lambda y_3}{\nu + \lambda}\right)$$

and *M* divides *CA* in the ratio $\lambda : \nu$. Similarly *N* divides *AB* in the ratio $\mu : \lambda$.

Thus we obtain Ceva's theorem that, when *AL*, *BM*, *CN* are concurrent at *P*, the product of the ratios in which the sides *AB*, *BC*, *CA* are divided at *L*, *N*, *M* is $+1$.

Now the ratios $\lambda : \mu : \nu$ uniquely define the point *P* and the point *P* uniquely defines the ratios $\lambda : \mu : \nu$. The set of three numbers (λ, μ, ν) are called the *homogeneous co-ordinates* of the point *P*. It is only the ratio of the co-ordinates which is significant, and $(k\lambda, k\mu, k\nu)$ must be taken to represent the same point.

Equations in homogeneous co-ordinates must contain only terms which are of the same degree. Thus the general first degree equation is

$$a\lambda + b\mu + c\nu = 0,$$

and it can be shown that this represents the general straight line. The second degree equation represents the general conic.

If for fixed k_1, k_2, k_3

$$\xi = k_1 \lambda, \quad \eta = k_2 \mu, \quad \zeta = k_3 \nu$$

then (ξ, η, ζ) represent the most general form of homogeneous co-ordinates used. Cartesian co-ordinates may be regarded as a particular case, when the two points B and C tend to infinity and, in order to keep the co-ordinates finite, k_2 and k_3 tend to zero.

The group of transformations which represent the change from one set of rectangular Cartesian co-ordinates to another represents the group of motions of a rigid body, since a rigid body could be moved from coincidence with one set of axes into coincidence with another. The measurements of length, angle and area are invariants of this group of rotations of a rigid body. Hence, in general, if we wish to study the geometrical properties concerning length, angles, areas, the system of Rectangular Cartesian co-ordinates is the most convenient to use.

The group of transformations from one set of homogeneous co-ordinates to another, however, is called the full linear group, and each of the three co-ordinates may be replaced by any linear combination of them, provided that the co-ordinates remain independent. Thus

$$\begin{aligned} \xi' &= l_1 \xi + m_1 \eta + n_1 \zeta, \\ \eta' &= l_2 \xi + m_2 \eta + n_2 \zeta, \\ \zeta' &= l_3 \xi + m_3 \eta + n_3 \zeta. \end{aligned}$$

The invariants under this system of transformations are quite different and are called projective invariants.

Length is not invariant, nor is the ratio in which a line AC is divided at B. But if $ABCD$ are four points on a straight line, then the *cross ratio*

$$\frac{AB.\ CD}{AD.\ BC}$$

is invariant under the group of transformation. Cross ratios are used in projective geometry as freely as lengths are used in Euclidian geometry.

Consider two planes P_1 and P_2 intersecting in a line L. Let O be any point not in either plane. Then if C_1 is any curve in P_1 suppose that O is joined to each point on the curve C_1 and produced until it cuts the plane P_2 in a corresponding

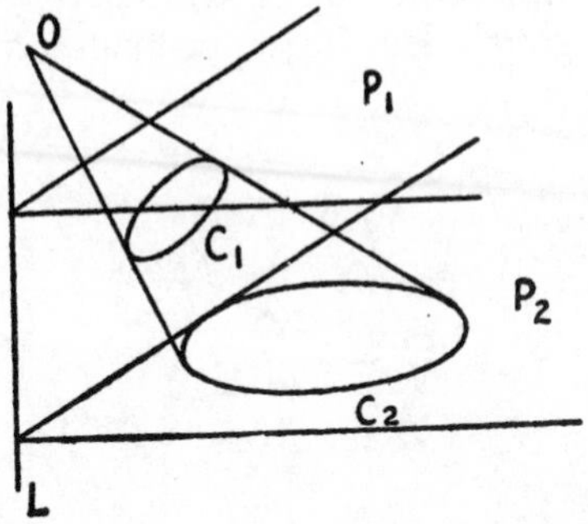

point. Then the corresponding points form a curve C_2 in the plane P_2 which is called the *projection* of the curve C_1 from O on to P_2.

Certain properties are unchanged in this projection, and these are the projective invariants. Thus a line becomes a line, a conic becomes a conic, and cross ratios are invariant. For such geometric properties as are invariant under this process of projection, the use of homogeneous co-ordinates is the appropriate method.

CHAPTER XI

MATRICES AND DETERMINANTS

CONSIDERING the case of Cartesian co-ordinates, not necessarily orthogonal, let the co-ordinates of a point P be

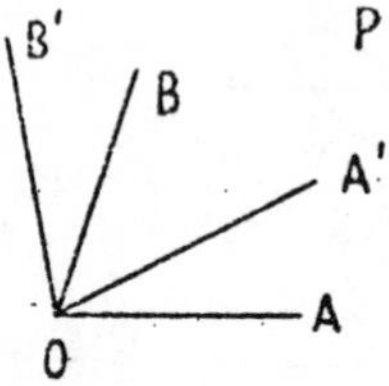

(x, y) so that

$$OP = x\,OA + y\,OB.$$

Referred to a different set of co-ordinates, but with the same origin O let the co-ordinates be (x', y') so that

$$OP = x'\,OA' + y'\,OB'.$$

Now suppose that relative to the new axes the co-ordinates of A, B are respectively (a, b) and (c, d) so that

$$OA = a\,OA' + b\,OB', \quad OB = c\,OA' + d\,OB'.$$

Then $OP = x\,(a\,OA' + b\,OB') + y\,(c\,OA' + d\,OB')$

$$= (a\,x + c\,y)\,OA' + (b\,x + d\,y)\,OB'$$

$$= x'\,OA' + y'\,OB'.$$

Hence

$$x' = a\,x + c\,y,$$
$$y' = b\,x + d\,y.$$

This pair of equations expresses the new co-ordinates (x', y') in terms of the old (x, y), and is called a *linear transformation.*

The linear transformation is specified by the four coefficients a, b, c, d, which are written down in a square array according to their position in the equations, thus

$$A = \begin{bmatrix} a, & c \\ b, & d \end{bmatrix}.$$

This array is called a matrix and we denote it by the capital letter A.

The resultant of two linear transformations taken consecutively is a third transformation. Thus if

$$x' = ax + cy, \quad x'' = a'x' + c'y',$$
$$y' = bx + dy, \quad y'' = b'x' + d'y',$$

then

$$x'' = (a'a + c'b)x + (a'c + c'd)y$$
$$y'' = (b'a + d'b)x + (b'c + d'd)y.$$

The matrix of the resultant is defined as the product of the matrices of the separate transformations. We write

$$\begin{bmatrix} a', & c' \\ b', & d' \end{bmatrix} \begin{bmatrix} a, & c \\ b, & d \end{bmatrix} = \begin{bmatrix} a'a + c'b, & a'c + c'd \\ b'a + d'b, & b'c + d'd \end{bmatrix}.$$

The order of the matrices on the left-hand side of the equation is important. The transformation corresponding to the matrix on the right must be taken before the transformation corresponding to the matrix on the left. If the order of the matrices were reversed a different product would be obtained. Matrices are not commutative.

The rule for multiplying matrices will be observed. Take any row of the left-hand matrix and any column of the right, multiply corresponding terms and add. The result goes into the product matrix in the position of the row and column chosen. A similar result holds for matrices with three or more rows and columns.

The multiplication rule can be extended to rectangular matrices, that is, when the number of rows and columns in a matrix are not necessarily equal, provided that the number of columns in the left-hand matrix is equal to the number of rows in the right-hand matrix.

The linear transformation can then be expressed in matrix form. Putting

$$X = \begin{bmatrix} x \\ y \end{bmatrix}, \quad X' = \begin{bmatrix} x' \\ y' \end{bmatrix},$$

then

$$X' = AX$$

is equivalent to the two equations of the transformation.

The *transpose* of a matrix is the matrix obtained by interchanging the rows with the columns. Transposition is indicated by putting the sign $\sim$ above the matrix. Thus

$$\tilde{X} = [x, y], \ \tilde{X}' = [x', y'],$$

$$\tilde{A} = \begin{bmatrix} a, & b \\ c, & d \end{bmatrix}.$$

If matrices are transposed, the order of the multiplication must be reversed. Thus if

$$C = A\,B,$$

then

$$\tilde{C} = \tilde{B}\,\tilde{A}.$$

If $X' = A\,X$ where

$$A = \begin{bmatrix} a, & c \\ b, & d \end{bmatrix}$$

then in general x and y can be expressed back in terms of x' and y' in the form

$$X = A'\,X',$$

by solving for x and y the equations

$$a\,x + c\,y = x',$$
$$b\,x + d\,y = y'.$$

The matrix A' is called the reciprocal of A and is denoted by A^{-1}.

This solution fails, however, if $(b\,x + d\,y)$ is multiple of $(a\,x + c\,y)$, since on attempting to solve the equations, the elimination of x eliminates also y from the equations at the same time. Thus if

$$x + 2\,y = x',$$
$$2\,x + 4\,y = y',$$

then there is no solution for x, y in terms of x' and y'. Instead we obtain the relation

$$2\,x' - y' = 0.$$

The matrix of coefficients

$$\begin{bmatrix} 1, & 2 \\ 2, & 4 \end{bmatrix}$$

is called a *singular* matrix. It has no reciprocal. The condition that a two-rowed matrix

$$\begin{bmatrix} a & c \\ b & d \end{bmatrix}$$

should be singular is that the rows should be dependent on one another. That is to say, there is a relation such that λ_1

times the first row $+ \lambda_2$ times the second row is identically zero, i.e.

$$[\lambda_1, \lambda_2] \begin{bmatrix} a & c \\ b & d \end{bmatrix} = 0.$$

A singular matrix thus has a left-hand *factor of zero*. Similarly there is a right-hand factor of zero

$$\begin{bmatrix} a & c \\ b & d \end{bmatrix} \begin{bmatrix} p_1 \\ p_2 \end{bmatrix} = 0.$$

The condition for the matrix to be singular may be expressed in terms of the elements of the matrix in the form

$$a\,d - b\,c = 0.$$

The expression $a\,d - b\,c$ is called the *determinant* of the matrix and it is written

$$\begin{vmatrix} a & c \\ b & d \end{vmatrix} = a\,d - b\,c.$$

Similarly for three-rowed matrices. The matrix

$$\begin{bmatrix} a_1, & b_1, & c_1 \\ a_2, & b_2, & c_2 \\ a_3, & b_3, & c_3 \end{bmatrix}$$

is called singular if there exists a left-hand factor of zero, which implies also a right-hand factor of zero. Thus

$$[\lambda, \mu, \nu] \begin{bmatrix} a_1 & b_1 & c_1 \\ a_2 & b_2 & c_2 \\ a_3 & b_3 & c_3 \end{bmatrix} = 0.$$

The function of the elements which vanishes when the matrix is singular is called the *determinant*, and for a three-rowed matrix it is

$$\begin{vmatrix} a_1, & b_1, & c_1 \\ a_2, & b_2, & c_2 \\ a_3, & b_3, & c_3 \end{vmatrix} = \begin{array}{l} a_1b_2c_3 + a_2b_3c_1 + a_3b_1c_2 \\ - a_1b_3c_2 - a_2b_1c_3 - a_3b_2c_1. \end{array}$$

All the products of three elements of the matrix, one from each row, one from each column, occur in the expansion of the determinant, with coefficients $+1$ or -1. The terms from the leading diagonal from the top left to the bottom right corner form a product with coefficient $+1$. From this, an interchange of two rows gives a product with coefficient -1; a second interchange gives coefficient $+1$, and so on.

Similarly the determinant of a four-rowed matrix is

defined with 24 terms, of which 12 are positive, 12 negative.

For square matrices of any order, if the determinant is zero, there is no reciprocal, but there exist both right-hand and left-hand factors of zero.

The matrix corresponding to the identical transformation with unity in each position in the leading diagonal and zero elsewhere is denoted by I and plays the role of the number 1 in arithmetic.

If T is a non-singular matrix, then the matrix $T^{-1}A\,T$ has many properties similar to those of A, and is called the transform of A by the matrix T. In general, a matrix can be transformed into a matrix in which the only non-zero terms are in the leading diagonal, and this is called a diagonal matrix. It is accomplished as follows. Suppose that A is a three-rowed matrix.

If a one column matrix X can be found such that

$$A\,X = \lambda\,X$$

where λ is a number, then X is called a *pole* of the matrix, and λ the corresponding *latent root*. For this to be the case we must have

$$[A - \lambda\,I]\,X = 0,$$

and thus $A - \lambda\,I$ is a singular matrix and its determinant is zero.

Taking

$$A = \begin{bmatrix} a_1, & b_1, & c_1 \\ a_2, & b_2, & c_2 \\ a_3, & b_3, & c_3 \end{bmatrix}$$

$$|A - \lambda\,I| = \begin{vmatrix} a_1 - \lambda & b_1 & c_1 \\ a_2 & b_2 - \lambda & c_2 \\ a_3 & b_3 & c_3 - \lambda \end{vmatrix}$$

$$= -\lambda^3 + (a_1 + b_2 + c_3)\,\lambda^2 - (a_1 b_2 - a_2 b_1 + a_1 c_3 - a_3 c_1 + b_2 c_3 - b_3 c_2)\,\lambda + |A| = 0.$$

This is called the characteristic equation of the matrix. It is a cubic equation which in general will have three roots $\lambda_1, \lambda_2, \lambda_3$, each of which is a latent root of the matrix. Corresponding to each latent root, $A - \lambda\,I$ is singular and has a right-hand factor of zero which is the corresponding pole. Suppose that

$$A\,X_1 = \lambda_1\,X_1,\; A\,X_2 = \lambda_2\,X_2,\; A\,X_3 = \lambda_3\,X_3.$$

Then putting the three poles X_1, X_2, X_3 together to form the three columns of a matrix T, it follows that

$$A\,T = T \begin{bmatrix} \lambda_1 & & \\ & \lambda_2 & \\ & & \lambda_3 \end{bmatrix}$$

and $T^{-1}\,A\,T$ is the diagonal matrix whose diagonal elements are the latent roots.

If it should happen that two of the roots of the characteristic equation were equal, this procedure might fail. Thus the matrix $\begin{bmatrix} 1, & 1 \\ 0, & 1 \end{bmatrix}$ cannot be transformed to diagonal form.

There are some special types of matrix. A *symmetric* matrix is one which satisfies $\tilde{A} = A$. It can be shown the latent roots of a real symmetric matrix are all real numbers, never complex numbers, and even when two roots are equal it can still be transformed into diagonal form.

A one-column matrix $X = \begin{bmatrix} x_1 \\ x_2 \\ x_3 \end{bmatrix}$ is called a *vector* and the *norm* of the vector is $\tilde{X}\,X = x_1^2 + x_2^2 + x_3^2$.

An *orthogonal* matrix is one which leaves the norm of every vector on which it operates invariant. If A is orthogonal and $X' = A\,X$, then

$$\tilde{X}\,X = \tilde{X}'\,X' = \tilde{X}'\,\tilde{A}\,A\,X.$$

This will be satisfied for all vectors X if, and only if,

$$\tilde{A}\,A = I,$$

which is the condition for an orthogonal matrix. If

$$A = \begin{bmatrix} a_1, & b_1, & c_1 \\ a_2, & b_2, & c_2 \\ a_3, & b_3, & c_3 \end{bmatrix}$$

then the condition is

$$\tilde{A}\,A = \begin{bmatrix} a_1, & a_2, & a_3 \\ b_1, & b_2, & b_3 \\ c_1, & c_2, & c_3 \end{bmatrix} \begin{bmatrix} a_1 & b_1 & c_1 \\ a_2 & b_2 & c_2 \\ a_3 & b_3 & c_3 \end{bmatrix} = \begin{bmatrix} 1 & & \\ & 1 & \\ & & 1 \end{bmatrix}$$

thus $a_1^2 + a_2^2 + a_3^2 = b_1^2 + b_2^2 + b_3^2 = c_1^2 + c_2^2 + c_3^2 = 1$,
and $a_1\,b_1 + a_2\,b_2 + a_3\,b_3 = a_1\,c_1 + a_2\,c_2 + a_3\,c_3 =$

$$b_1\,c_1 + b_2\,c_2 + b_3\,c_3 = 0.$$

Similar results hold for the columns. The importance of orthogonal matrices lies in the fact that they represent a transformation from one set of rectangular Cartesian co-ordinates to another, provided that the origin is unchanged.

It can be shown that an orthogonal matrix may always be chosen to transform a symmetric matrix into diagonal form.

These matrices have immediate application to algebraic geometry. The equation to a line

$$l x + m y + n = 0$$

may be expressed in matrix form as

$$\tilde{L} X = 0$$

where $\tilde{L} = [l, m, n]$ and $X = \begin{bmatrix} x \\ y \\ 1 \end{bmatrix}$

Then if a change of co-ordinates is made, this can be expressed as $X' = A X$, whence the equation to the line becomes

$$\tilde{L} X = \tilde{L} A^{-1} X' = 0,$$

which is the same as $\tilde{L}' X' = 0$, if

$$\tilde{L}' = \tilde{L} A^{-1}, \; L' = \tilde{A}^{-1} L.$$

The set of coefficients l, m, n is transformed by the matrix $\tilde{A}^{-1}$.

The second degree curve

$$a x^2 + b y^2 + 2 h x y + 2 f y + 2 g x + c = 0$$

can be expressed as $\tilde{X} \phi X = 0$, where ϕ denotes the symmetric matrix

$$\begin{bmatrix} a, & h, & g \\ h, & b, & f \\ g, & f, & c \end{bmatrix}$$

If X_1 corresponds to a point on the curve, then the tangent at the point has equation $\tilde{X}_1 \phi X = 0$. If the equation should represent two lines, then at the point of intersection the tangent is indeterminate and its equation must be satisfied identically, so that

$$\tilde{X}_1 \phi = 0.$$

This can happen if, and only if, the matrix ϕ is singular. Hence the condition for two lines is that

$$|\phi| = abc + 2fgh - af^2 - bg^2 - ch^2 = 0.$$

To transform the equation to its canonical form, it is convenient to separate the translations of the axes from the rotations. The translations leave the second degree terms with coefficients a, b, h unchanged. The matrix corresponding to these terms is:

$$\bar{\phi} = \begin{bmatrix} a, & h \\ h, & b \end{bmatrix}$$

This is a symmetric matrix which can be transformed into diagonal form by an orthogonal matrix. The diagonal terms will be the roots λ_1, λ_2 of the characteristic equation

$$\lambda^2 - (a + b)\lambda + ab - h^2 = 0.$$

Corresponding to this orthogonal matrix there is a rotation of axes which brings the second degree terms to the form

$$\lambda_1 x^2 + \lambda_2 y^2.$$

The poles of the matrix ϕ give the directions of the new axes.

If neither λ_1 nor λ_2 is zero, that is, if $ab - h^2$ differs from zero, then a translation will bring to zero the terms $2gx$ and $2fy$, to give the canonical form of an ellipse or hyperbola.

If $ab - h^2 = 0$, we can take $\lambda_1 = 0$. A translation will then reduce to zero the term $2fy$ and the independent term, to give the canonical form of the parabola.

The same method can be applied to second degree surfaces in three dimensions.

For homogeneous co-ordinates and projective geometry, the matrix of transformation is the general non-singular matrix. The matrix method can be applied with equal effectiveness.

CHAPTER XII

INVARIANTS AND TENSORS

THE last two chapters have served to indicate the importance of the concept of invariants in algebraic geometry. But not alone in algebraic geometry, also in applied mathematics and mathematical physics the method of invariants is fundamental. The instance of the special theory of relativity will illustrate. But before discussing this, a formal definition of invariants and other concomitants will be opportune.

Suppose that there is given a set of n variables, x_1, x_2, ... , x_n, which may represent the co-ordinates of a point in two, three or more dimensions, which may be transformed by the linear substitutions of group. This would correspond to the transition from one co-ordinate system to another. It is supposed that there are given a set of algebraic forms in the variables, say $f(a_1, \ldots, a_m, x_1, \ldots, x_n)$ which is written concisely $f(a_i, x_i)$. $g\ (b_i, x_i)$, $h\ (c_i, x_i)$, etc., are similar forms. These are called the ground forms.

When the variables are subjected to a linear transformation of the group, the algebraic form will become an algebraic form of the same kind but with different coefficients, say $f\,(a_i, x_i) = f\,(a_i', x_i')$.

Then if there is a function of the coefficients $\phi\,(a_i, b_i, c_i)$ which after transformation becomes equal to the same function of the transformed coefficients, this function is called an *absolute invariant* under the group of transformations.

When transformations are allowed for which the determinant differs from unity, it is more usual to consider *relative invariants* for which the transformed function is a multiple of the original function by a factor which is some power of the determinant of the transformation. Using Δ to denote this determinant, the equation is

$$\phi\,(a_i, b_i, c_i) = \Delta^k\,\phi\,(a_i', b_i', c_i').$$

If instead there is a function which involves not only the coefficients, but also the co-ordinates of an arbitrary point

$$\phi(a_i, b_i, c_i;\ x_i) = \Delta^k \phi(a_i', b_i', c_i';\ x_i'),$$

this is called a *covariant*.

There are also functions which involve the coefficients and the co-ordinates of two or more points, which are called *mixed concomitants*. All such invariants, covariants and mixed concomitants are the *concomitants* of the ground forms under the group of transformations.

Turning now to the special theory of relativity, the manner of the discovery was as follows.

Since the earth rotates about its axis once a day, a point on the earth's surface at the equator must be moving with a speed of about a thousand miles an hour as compared with the centre of the earth. Further, because of the motion of the earth round the sun, the earth itself must be moving with a velocity of about sixty thousand miles an hour relative to the sun. We cannot perceive these velocities, as all bodies in the neighbourhood are moving together, and the equations of mechanics are such as to remain invariant when all bodies are given a simultaneous relative velocity.

It is known, however, that light travels with a speed of some 186,000 miles a second, the same in all directions. But this is something which should be modified by the relative velocity of the earth. Owing to the intrinsic velocity of the earth and the apparatus used, the apparent velocity of light ought to be different according as the light is moving with, across or against this velocity. Michelson and Morley designed a very sensitive apparatus called an interferometer which could detect extremely small changes in the velocity of light, expecting that it would thus reveal the absolute velocity of the apparatus. The surprising result was, however, that no change at all could be detected in the apparent velocity of light. The apparatus behaved exactly as if it was absolutely at rest.

Only one explanation is possible. The traditional equations based on Newton's theory of mechanics, showing the effects of relative velocity on a body, must be incorrect. They must be replaced by another group of transformations which leave the velocity of light invariant.

Taking rectangular Cartesian co-ordinates x, y, z in space and using c to denote the velocity of light, a pulse of light travelling outwards from the origin will reach the point (x, y, z) after a time t if $x^2 + y^2 + z^2 = c^2 t^2$. Hence $x^2 + y^2 + 2^2 - c^2 t^2$ must remain invariant for the appropriate group of transformations.

In the xy — plane, the transformations which leave $x^2 + y^2$ invariant are well known to be the orthogonal transformations such as

$$x' = \frac{x + ky}{\sqrt{(1 + k^2)}}, \quad y' = \frac{y - kx}{\sqrt{(1 + k^2)}}.$$

In the xt — plane where y^2 is replaced by $-c^2 t^2$, the following modification is required:

$$x' = \frac{x + kct}{\sqrt{(1 - k^2)}}, \quad ct' = \frac{ct + kx}{\sqrt{(1 - k^2)}}.$$

These are the appropriate equations which connect the measurements made from a moving body with those made from a stationary body.

The principal differences from the traditional theory are these. The concept of time is not the same for a moving body. If two events in different places are reckoned to be simultaneous as observed from the earth, then as observed from an express train, one may appear to be earlier than the other. Also, when a rigid body is given a velocity, it must experience a contraction of its length in the direction of its motion. This contraction, called after the mathematician who investigated this effect the *Fitzgerald contraction*, is extremely small in connection with the velocities normally encountered, because the velocity of light is so large, and this explains why it had previously escaped detection. The last effect is that a clock, when moving, goes slower than when it is still.

These equations expressing the effect of relative velocities are called a Lorentz transformation, after another mathematician intimately connected with the discovery.

If an aeroplane moving with velocity u fires a bullet forwards with velocity v relative to the aeroplane, the velocity of the bullet relative to the earth will not be $u + v$ but $(u + v)/\left(1 + \frac{uv}{c^2}\right)$, which is slightly less. A force which would cause an acceleration f for a stationary body would thus produce a slightly smaller acceleration in a moving body. This is equivalent to an increase in the inertial mass of a body. Einstein recast the equations of mechanics so as to make these invariant for a Lorentz transformation, and one of his conclusions was that if the mass of a body at rest is m, then the mass when it is moving with velocity v must be

$$m/\sqrt{\left(1 - \frac{v^2}{c^2}\right)}.$$

Further, the kinetic energy, or energy of motion, which was traditionally equated to $\frac{1}{2} m v^2$, must be expressed by the function $m c^2/\sqrt{\left(1 - \frac{v^2}{c^2}\right)}$. To a first approximation when v is small, this gives $m c^2 + \frac{1}{2} m v^2$. The first term $m c^2$ is constant and is not observed as available energy; the second term gives the traditional expression as an approximation.

This identification of mass with energy is of extreme importance in modern physics, especially concerning nuclear disintegration, where the difference in mass of atoms before and after disintegration is a measure of the available energy.

For the equations of physics to be correct, they must be expressed in such a form that they are invariant under the Lorentz transformations. Hence all the effects must be expressible as concomitants under the Lorentz group. For this reason, in physics also, it is of great importance to be able to write down the concomitants of a system of ground forms, as it is in terms of these alone that the equations of physics must be written.

A very effective device for the construction of concomitants is the *tensor*, which will now be described.

Suppose that the set of variables which are transformed by the basic transformation is $x^1, x^2, \ldots\ldots, x^n$. The suffixes are written at the top instead of at the bottom for reasons that will become apparent. The variables are said to form a *contragredient tensor of rank* one. If $\Sigma u_i x^i$ is a linear form which remains invariant, then when the x^i's undergo a transformation the u_i's must be made to undergo the reciprocal transformation. The quantities u_i are said to form a *cogredient* tensor of rank one. Upper suffixes are used for contragredient and lower suffixes for cogredient tensors. It is usual to omit the Σ or summation sign and make the convention that whenever the same suffix is used for an upper and for a lower suffix, a summation is made for all values of this suffix. Thus $u_i x^i$ denotes $\Sigma u_i x^i$. This process is called contraction.

A quadratic form in the n variables could be written as $a_{ij} x^i x^j$. When the x^i's undergo a transformation, the coefficients a_{ij} will undergo the reciprocal transformation with respect to each of the suffixes i, j. They are said to form a cogredient tensor of rank 2. Tensors of any rank are similarly defined.

A tensor may have both upper and lower suffixes, in which case it is called a *mixed tensor*.

The essential point is that, firstly, the product of any two tensors is also a tensor, and secondly any mixed tensor can be contracted so as to eliminate one of the upper and one of the lower suffixes, and the result will be a tensor of lower rank. A tensor of rank zero is always an invariant.

Given a set of ground forms with the coefficients in tensor form, the result of multiplying the tensors in any way and contracting so as to form a tensor of zero rank, will always be a concomitant of the forms. If the variable tensor x^i is not introduced an invariant is obtained. If x^i is introduced the result is a covariant.

Under the orthogonal group, there is a metric form, that is to say a quadratic form $g_{ij} x^i x^j$ is left invariant. For the ordinary orthogonal group this is taken to be the sum of the squares of the variables, so that

$$g_{11} = g_{22} = g_{33} = \ldots . = g_{nn} = 1, \; g_{ij} = 0, \; (i \neq j).$$

But the theory is equally applicable to the Lorentz group. Taking x^1, x^2, x^3 for the space co-ordinates and x^0 for $c\,t$, the metric is given by

$$g_{11} = g_{22} = g_{33} = 1, \; g_{00} = -1, \; g_{ij} = 0, \; (i \neq j).$$

The metric tensor can be used with the ground form tensors for constructing concomitants.

There is also the alternating tensor which is of the same rank as the number of variables. Thus in four variables it satisfies

$$\Gamma_{ijkp} = 0 \text{ if two suffixes are equal,}$$

$\Gamma_{ijkp} = 1$ if $ijkp$ is obtained from 1, 2, 3, 4 by making an even number of interchanges.

$\Gamma_{ijkp} = -1$, for an odd number of interchanges.

This tensor can be either cogredient or contragredient.

The invariants of the quadratic form in three variables $a_{ij}\, x^i\, x^j$, subject to the metric $g_{ij}\, x^i\, x^j$, can be obtained as follows:

Firstly, there is also a tensor g^{ij} obtained from g_{ij} as follows:

$$g^{ij} = \Gamma^{ii'i''}\, \Gamma^{jj'j''}\, g_{i'j''}\, g_{i''j''}.$$

For the metric considered it takes the same values as g_{ij}.

Then

$$g^{ij}\, a_{ij}$$

is an invariant, as is also

$$g^{ij}\, g^{kp}\, a_{ik}\, a_{jp}.$$

A third invariant is

$$\Gamma^{ijk}\, \Gamma^{pqr}\, a_{ip}\, a_{jq}\, a_{kr}.$$

If new variables are introduced such that these are cogredient instead of contragredient, thus u_i, making $u_i\, x^i$ invariant, then expressions involving these are called contravariants.

The above quadratic has a contravariant

$$\Gamma^{ijk}\, \Gamma^{pqr}\, a_{ip}\, a_{jq}\, u_k\, u_r.$$

These concomitants are very similar to those obtained in a previous chapter concerning the second degree terms of the conic.

It is quite usual in physical theory to express equations in terms of tensors, thus making sure that the resulting expressions will be invariant for the appropriate group of transformations, usually the Lorentz group.

The chief difficulty with tensors is that an almost unlimited number of expressions with products and contractions can be written down. These are not all independent concomitants, however. Many expressions will reduce to zero when evaluated, and many quite different expressions represent the same thing. A much deeper theory is necessary to determine which are the independent concomitants.

It is necessary to distinguish between symmetric and antisymmetric tensors. A tensor of rank 2, a_{ij} is symmetric if

$a_{ij} = a_{ji}$. It is anti-symmetric if $a_{ij} = -a_{ji}$. Any tensor of rank 2 can be expressed as a sum of symmetric and an anti-symmetric tensor. Thus

$$a_{ij} = \tfrac{1}{2}(a_{ij} + a_{ji}) + \tfrac{1}{2}(a_{ij} - a_{ji}).$$

The symmetric and the anti-symmetric parts transform entirely independently of one another.

A tensor of rank 3 separates into three parts, a symmetric tensor such that

$$a_{ijk} = a_{ikj} = a_{jik} = a_{jki} = a_{kij} = a_{kji},$$

also an anti-symmetric tensor such that

$$a_{ijk} = a_{jki} = a_{kij} = -a_{ikj} = -a_{jik} = -a_{kji},$$

and a third tensor that is not so simple, but is partly symmetric and partly anti-symmetric.

An understanding of the nature of these tensors of higher rank can only be obtained by methods of substitutional analysis or group representational methods, which will be mentioned in a later chapter.

CHAPTER XIII

ALGEBRAS

ONE of the earliest and perhaps the most spectacular discovery of a non-commutative algebra was the discovery of *Quaternions* by Hamilton early last century.

In three-dimensional space it is convenient to treat quantities which have both magnitude and direction, such quantities as forces, velocities, displacements, electrical intensities, as *vectors* and to represent them by a directed line. Addition of vectors is defined by means of the parallelogram law.

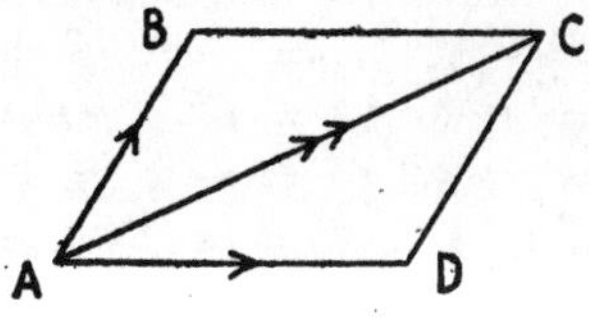

Thus, in the diagram, $ABCD$ being a parallelogram,

$$AB + AD = AC.$$

To complete the analogy with number systems, some analogue of multiplication is needed. But the position in respect of multiplication is not so simple. Some properties which could be associated with a product of the two vectors, such as the area of the parallelogram $ABCD$, when AB and AD are the vectors, is equal to zero when the vectors are in the same direction, and is a maximum for given magnitudes of the vectors, when the directions are perpendicular. Such a product, it is convenient to treat as a vector, associating it with the direction perpendicular to the plane of the parallelogram.

Another interpretation of multiplication could occur when the vector AB represents a force and the vector AD a displacement from A to D of the point of application. Then if the work done by the force in the displacement is regarded as the product, this is a maximum when AB and AD are in the same direction, and is zero when they are perpendicular. Such a product is not associated with any direction and is purely numerical. It is called a *scalar product*. The other product is called the *vector product*.

Hamilton discovered that by putting together the scalar and the vector products to produce a system with four components, the scalar and the three components of the vector, then this system showed elegant analogies with ordinary number systems, but differed from those notably in that the commutative law of multiplication was not obeyed.

Taking rectangular Cartesian co-ordinates (x, y, z) and denoting a vector of unit magnitude parallel to OX by i, parallel to OY by j, and parallel to OZ by k, then a vector with components (x, y, z) would be denoted by $x\,i + y\,j + z\,k$.

If for the law of multiplication it is assumed that

$$i^2 = -1, \quad j^2 = -1, \quad k^2 = -1,$$
$$ij = k, \quad jk = i, \quad ki = j,$$
$$ji = -k, \quad kj = -i, \quad ik = -j,$$

then the product of two vectors $(x\,i, y\,j, z\,k)\,(x'\,i, y'\,j, z'\,k)$ will have a scalar part which is equal to the scalar product with the sign reversed, and a vector part which is equal to the vector product. Thus

$$\begin{aligned}&(x\,i + y\,j + z\,k)\,(x'\,i + y'\,j + z'\,k)\\ &= -(x\,x' + y\,y' + z\,z') + i\,(y\,z' - y'\,z) + j\,(z\,x' - z'\,x)\\ &\quad + k\,(x\,y' - x'\,y).\end{aligned}$$

The *quaternions* so formed obey all the laws of numbers except for the commutative law of multiplication. In particular, and most important, they are associative, i.e. three quaternions A, B, C satisfy

$$(A\,B)\,C = A\,(B\,C).$$

If $Q = x_0 + i\,x_1 + j\,x_2 + k\,x_3$ is any quaternion, the quaternion obtained by changing the sign of the vector part, $\bar{Q} = x_0 - i\,x_1 - j\,x_2 - k\,x_3$, is called the conjugate quaternion. The product of a quaternion with its conjugate is a positive non-zero scalar

$$(x_0 + i\,x_1 + j\,x_2 + k\,x_3)\,(x_0 - i\,x_1 - j\,x_2 - k\,x_3) = x_0^2 + x_1^2 + x_2^2 + x_3^2,$$

which is called the *norm* of the quaternion. It is thus always possible to divide by a quaternion, except of course by zero.

$$\frac{1}{x_1 + i\,x_1 + j\,x_2 + k\,x_3} = \frac{x_0 - i\,x_1 - j\,x_2 - k\,x_3}{x_0^2 + x_1^2 + x_2^2 + x_3^2}$$

Quaternions with real coefficients, or *real* quaternions, are thus said to form a *division algebra.*

If complex coefficients are allowed, however, division may fail, for with complex numbers it is possible that $x_0^2 + x_1^2 + x_2^2 + x_3^2 = 0$ without all the quantities x_0, x_1, x_2, x_3 being zero. Complex quaternions do *not* form a division algebra. It can be shown that they are equivalent to the algebra of two-rowed matrices if we set

$$1 = \begin{bmatrix} 1 & 0 \\ 0 & 1 \end{bmatrix},\ i = \begin{bmatrix} \sqrt{-1} & 0 \\ 0 & -\sqrt{-1} \end{bmatrix},\ j = \begin{bmatrix} & 1 \\ -1 & \end{bmatrix},$$

$$k = \begin{bmatrix} 0 & \sqrt{-1} \\ \sqrt{-1} & 0 \end{bmatrix}.$$

An alternative system of space analysis is due to Grassman. It concerns only vector products, but is applicable to any number of dimensions, whereas quaternions are confined to three.

In Grassman's space analysis a vector with components $x_1, x_2, \ldots, x_n$ is associated with $e_1 x_1 + e_2 x_2 + \ldots + e_n x_n$ where the quantities $e_1, e_2, \ldots, e_n$ satisfy the following equations:

$$e_i^2 = 0,\ e_i e_j = -\,e_j e_i.$$

In three dimensions, a vector would be of the form $a\,e_1 + b\,e_2 + c\,e_3$. An element of area would take the form $a\,e_2 e_3 + b\,e_3 e_1 + c\,e_1 e_2$, and would thus be distinguishable from the vector with which it would be identified in quaternions. An element of volume would have the single component $k\,e_1 e_2 e_3$, for $e_1 e_2 e_3 = e_2 e_3 e_1 = e_3 e_1 e_2 = -\,e_1 e_3 e_2 = -\,e_2 e_1 e_3 = -\,e_3 e_2 e_1$.

The algebra yields immediately the formula for the area of a triangle as a half of the product of two sides, or the formula for the volume of a tetrahedron as one sixth of the product of three concurrent edges.

We are concerned here more with the algebras themselves than with the geometrical applications. The Grassman algebra differs very much from the quaternion algebra. With the real quaternion algebra there are no factors of zero; that is to say $x y = 0$ if, and only if, either $x = 0$ or $y = 0$. In the Grassman algebra every element is nilpotent, which is a way of saying that some power of the element is equal to zero. This must be the case, for every product of more than n elements corresponding to n dimensions must be zero. Actually the square of a vector $\Sigma x_i e_i$ is zero. But the square of every element of the algebra is not zero, for

$$(e_1 e_2 + e_3 e_4)^2 = 2 e_1 e_2 e_3 e_4.$$

The algebraic generalization of these algebras is as follows. It is supposed that scalar multiplication is possible, and hence we start with some field F from which the coefficients can be drawn.

The algebra is then defined as an aggregate of elements $a, b, c, \ldots$ subject to the operations of addition, multiplication and scalar multiplication which satisfy the following rules.

If a and b are elements, so is $(a + b)$ and addition is commutative and associative.

If a and b are elements, so is $a b$. If multiplication is associative it is called an associative algebra. It will be assumed that the algebras considered here are associative.

If a is an element and α is a member of the field F, then $\alpha a = a \alpha$ is also an element.

Multiplication and scalar multiplication are distributive with respect to addition.

The algebra is a finite algebra if a finite number of elements $e_1, e_2, \ldots, e_n$ can be found such that every element x of the algebra can be expressed in the form

$$x = \alpha_1 e_1 + \alpha_2 e_2 + \ldots . \alpha_n e_n$$

where $\alpha_1, \alpha_2, \ldots, \alpha_n$ belong to the field F.

If n is the least possible number of such elements the algebra is said to be of order n, and the elements $e_1, e_2, \ldots, e_n$ are said to form a basis of the algebra.

Starting from such a general definition, the properties of such algebras have been very closely examined, and the following is an account of the results obtained.

A subset B of elements from an algebra A which themselves form an algebra is called a *sub-algebra*. If when a belongs to A and b to B then $a b$ and $b a$ both belong to B, then B is an *invariant sub-algebra.* If an algebra A is the sum of two invariant sub-algebras with no common element, then A is *reducible* and equivalent to the direct sum of these sub-algebras. Clearly these sub-algebras are quite independent of one another and can be studied separately.

An element x is nilpotent if $x^n = 0$ for some n. x is properly nilpotent in A if $x\,y$ and hence also $y\,x$ is nilpotent for every element y of A. The properly nilpotent elements of an algebra A form an invariant sub-algebra.

Relative to this nilpotent invariant sub-algebra of A one can define the *difference algebra* A' which is effectively equivalent to equating to zero the properly nilpotent elements. The difference algebra possesses no properly nilpotent elements.

Then it can be shown that the algebra A' can be expressed as a direct sum of irreducible algebras, and each irreducible algebra is equivalent to the direct product of a matrix algebra and a division algebra. That is to say, each element can be regarded as a matrix of which the elements are not necessarily numbers, but elements of the division algebra. And all such matrices of elements of the division algebra are elements of the given irreducible algebra.

As regards the division algebras, the possible types of these depend upon the field over which the coefficients are taken. If the field is the complex numbers, then the only division algebra is the algebra of order one which consists of the complex numbers themselves. Over the real numbers there are exactly three division algebras, the real numbers themselves, an algebra of order two which is equivalent to the complex numbers, and real quaternions.

Over the rationals, or over an algebraic field, more complex division algebras are possible, which depend on the groups of automorphisms of an algebraic extension field.

One conclusion from the above is very significant. An algebra over the field of complex numbers containing no properly nilpotent element is equivalent to a direct sum of total matrix algebras. A total matrix algebra is just the complete algebra of matrices of order n^2.

This explains why quaternions, on the extension of the

field to include complex coefficients, become a two-rowed matrix algebra.

On the other hand, the Grassman algebra is completely eliminated when the nilpotent invariant sub-algebra is removed.

An important invariant property of an algebra concerns the *trace* of an element.

Suppose that an algebra of order n has basal elements $e_1, e_2, \ldots, e_n$, and that the multiplication table of these elements is given by

$$e_i\, e_j = \Sigma\, \Gamma_{ijk}\, e_k$$

where the coefficients Γ_{ijk} represent n^3 numbers of the field.

Then the *trace* of an element $x = \Sigma\, \xi_i\, e_i$ of the algebra is defined to be

$$\text{trace of } x = \sum_{ir} \xi_i\, \Gamma_{rir}.$$

Then it can be shown that the trace of the element x is independent of the choice of basal elements.

Hence if the algebra is expressed as a direct sum of invariant sub-algebras, the trace of an element of the algebra is equal to the sum of the traces of the corresponding elements in each sub-algebra.

If an algebra has an element m which satisfies $m\,a = a\,m = a$ for all elements a of the algebra, then m is called the modulus of the algebra. It plays exactly the role of the number 1. The trace of the modulus is easily shown, by taking it to be the first element e_1 of the basis, to be equal to the order of the algebra.

Further properties of the trace that are frequently employed in the study of algebras are as follows:

The trace of the sum of two elements is the sum of the traces.

The trace of a nilpotent element is zero.

If x is properly nilpotent, then the trace of every product $y\,x$ is zero.

These properties will be used in the next chapter in the study of group algebras.

CHAPTER XIV

GROUP ALGEBRAS

GROUPS and algebras have this in common, that they each employ a process of multiplication that is associative but not necessarily commutative. The problem that immediately suggests itself, then, is to examine the connection between the two theories. This connection is quite intimate, for connected with every finite group there is an associated algebra called the *group algebra*. It is sometimes called after Frobenius, who published a number of papers exploring this problem, the *Frobenius algebra* of the group.

If $S_1, S_2, \ldots, S_n$ are the elements of a group, then the algebra of order n with basal elements $e_1, e_2 \ldots, e_n$ and multiplication table $e_i e_j = e_k$ whenever $S_i S_j = S_k$ is called the group algebra.

No confusion arises if we use the same symbols S_i to represent the basal elements of the algebra e_i, as for the elements of the group, and we shall adopt this course. It is convenient to take the field of coefficients as the complex numbers.

Firstly, the group algebra possesses no properly nilpotent element. To show this we make use of the fact that the trace of a nilpotent element is zero.

Denote the identical element of the group by S_0, which will also be the modulus of the group algebra. $S_1, S_2, \ldots, S_{n-1}$ denote the other elements of the group.

It is clear that $S_r S_i = S_r$ if and only if $i = 0$. Hence the trace of S_i being equal to $\Sigma_r \Gamma_{rir}$ where $S_r S_i = \Sigma \Gamma_{rir} S_i$, it follows that the trace of S_0 is n, and the trace of every other group element in the algebra is zero.

Now let $x = \Sigma \xi_i S_i$ be any element of the group algebra. Then put $\bar{x} = \Sigma \bar{\xi}_i S_i^{-1}$, where $\bar{\xi}_i$ is the complex conjugate of ξ_i. The coefficient of the identical element S_0 in $\bar{x} x$ is clearly $\Sigma \bar{\xi}_i \xi_i$ which is definitely positive and cannot be zero unless $x = 0$. Hence the trace of $\bar{x} x$ is $n \Sigma \bar{\xi}_i \xi_i$ which is not zero, and $x x$ cannot be nilpotent. Thus x cannot be properly nilpotent, and the algebra has no properly nilpotent element.

The group algebra, being an algebra over the complex numbers possessing no properly nilpotent element, is thus equivalent to a direct sum of total matrix algebras.

Let C_ρ denote a class of the group. We also use C_ρ to denote the sum of the elements of the class considered as an element of the algebra. If T_i is an element of C_ρ and S any element of the group then $S^{-1} T_i S$ is also an element of C_ρ. Also $S^{-1} C_\rho S$ consists of the whole of the elements of the class C_ρ taken in a different order, and

$$S^{-1} C_\rho S = C_\rho$$
$$C_\rho S = S C_\rho.$$

Thus C_ρ commutes with every element of the group and also with every element of the group algebra. Conversely every element which commutes with every element of the algebra is a linear combination of the classes with scalar coefficients. Suppose that the classes are p in number.

Now consider the representations of the algebra as total matrix algebras. If an element x commutes with every element of the algebra, then the matrix representation in any sub-algebra must commute with every matrix in that sub-algebra. For this to be the case the representative matrix must be a numerical multiple of the identity matrix in that sub-algebra. The element x must thus be a linear combination of the moduli of the sub-algebras with numerical coefficients.

This implies that each class C_ρ can be expressed as a linear combination of the moduli of the sub-algebras, and each modulus of a sub-algebra can be expressed as a linear combination of the classes. For this to be the case there must be exactly the same number of matrix sub-algebras of the group as there are classes.

An example for the symmetric group on three symbols α, β, γ may perhaps make this more clear. There are three classes which may be expressed in terms of the group elements as follows:

$$C_0 = I,$$
$$C_1 = (\alpha\,\beta) + (\beta\,\gamma) + (\gamma\,\alpha),$$
$$C_2 = (\alpha\,\beta\,\gamma) + (\alpha\,\gamma\,\beta).$$

A basis for the group algebra may be taken as follows:

Put $x = \frac{1}{6}[I + (\alpha\,\beta\,\gamma) + (\alpha\,\gamma\,\beta) + (\alpha\,\beta) + (\beta\,\gamma) + (\gamma\,\alpha)],$

$$y = \tfrac{1}{6}[I + (\alpha\,\beta\,\gamma) + (\alpha\,\gamma\,\beta) - (\alpha\,\beta) - (\beta\,\gamma) - (\gamma\,\alpha)],$$
$$z_{11} = \tfrac{1}{3}[I + (\alpha\,\beta) - (\alpha\,\gamma) - (\alpha\,\beta\,\gamma)],$$
$$z_{22} = \tfrac{1}{3}[I + (\alpha\,\gamma) - (\alpha\,\beta) - (\alpha\,\gamma\,\beta)],$$
$$z_{12} = \tfrac{1}{3}[(\beta\,\gamma) - (\alpha\,\gamma) + (\alpha\,\gamma\,\beta) - (\alpha\,\beta\,\gamma)],$$
$$z_{21} = \tfrac{1}{3}[(\beta\,\gamma) - (\alpha\,\beta) + (\alpha\,\beta\,\gamma) - (\alpha\,\gamma\,\beta)].$$

Then it may be verified that

$$x^2 = x,\ y^2 = y,\ x\,y = y\,x = x\,z_{ij} = z_{ij}\,x = y\,z_{ij} = z_{ij}\,y = 0,$$
$$z_{11}^{\ 2} = z_{11},\ z_{22}^{\ 2} = z_{22},\ z_{12}\,z_{22} = z_{11}\,z_{12} = z_{12},\ z_{21}\,z_{11} =$$
$$z_{22}\,z_{21} = z_{21},\ z_{12}\,z_{12} = z_{12}\,z_{11} = z_{22}\,z_{12} = z_{22}\,z_{11} = 0,$$
$$z_{11}\,z_{22} = z_{11}\,z_{21} = z_{21}\,z_{21} = z_{21}\,z_{22} = 0.$$

The algebra is thus exhibited as a direct sum of three algebras, two unary algebras corresponding to x and y respectively, and a matrix algebra of degree 2.

The expressions of the moduli of the sub-algebras in terms of the classes are

$$x = \tfrac{1}{6}(C_0 + C_1 + C_2),$$
$$y = \tfrac{1}{6}(C_0 - C_1 + C_2),$$

and

$$z = z_{11} + z_{22}$$
$$= \tfrac{1}{3}[2\,I - (\alpha\,\beta\,\gamma) - (\alpha\,\gamma\,\beta)]$$
$$= \tfrac{2}{3}[C_0 - \tfrac{1}{2}C_2].$$

Conversely

$$C_0 = x + y + z,$$
$$C_1 = 3\,x - 3\,y,$$
$$C_2 = 2\,x + 2\,y - z.$$

The matrix representations of the group elements in the various sub-algebras are

$$I = x + y + \begin{bmatrix} 1 & \\ & 1 \end{bmatrix} z,$$

$$(\alpha\,\beta) = x - y + \begin{bmatrix} 1 & \\ -1 & -1 \end{bmatrix} z,$$

$$(\alpha\,\gamma) = x - y + \begin{bmatrix} -1 & -1 \\ & 1 \end{bmatrix} z,$$

$$(\beta\,\gamma) = x - y + \begin{bmatrix} & 1 \\ 1 & \end{bmatrix} z,$$

$$(\alpha\,\beta\,\gamma) = x + y + \begin{bmatrix} -1 & -1 \\ 1 & \end{bmatrix} z,$$

$$(\alpha\,\gamma\,\beta) = x + y + \begin{bmatrix} & 1 \\ -1 & -1 \end{bmatrix} z,$$

Each sub-algebra gives a matrix representation of the group. This is defined as a set of matrices M_i corresponding to the various elements of the group such that $M_i M_j = M_k$ whenever $S_i S_j = S_k$ is satisfied by the corresponding group elements.

In the general case if the group has p classes the sub-algebras of the group algebra give p separate matrix representations of the group. Of these, one will always be the representation for which each group element corresponds to unity.

Except for the representations of degree one, the representations are not uniquely defined by the group. For if in a representation M_i corresponds to the group element S_i, and T is a fixed matrix of the same degree, if

$$M_i M_j = M_k$$

then also

$$(T^{-1} M_i T)(T^{-1} M_j T) = T^{-1} M_k T.$$

Hence the set of matrices $T^{-1} M_i T$ also form a representation. The representations M_i and $T^{-1} M_i T$ are said to be equivalent.

Since the characteristic equation of a matrix is the same as that of a transform, this is the same for M_i as for $T^{-1} M_i T$. We choose from this characteristic equation the first coefficient, which is the sum of the leading diagonal terms in M_i, or the *spur* of the matrix. The set of spurs of the matrices M_i is called the *group character* and is used to specify the representation. The spur of M_i is called the *characteristic* of S_i. Conjugate elements of the group, being transforms of one another, have the same characteristics. The group character thus consists p distinct numbers, one for each class. The set of p characters corresponding to the distinct representations forms a square table called the table of characters.

For the symmetric group on three symbols, from the representations given above, we obtain the following characters:

Class . . .	(1^3)	$(1\ 2)$	(3)
Order . . .	1	3	2
Characters . .	1	1	1
	1	-1	1
	2	0	-1

The characters have curious orthogonal properties which can be proved by using the invariant trace of an element in an algebra. Thus in the above table, if the numbers in any row are squared, multiplied by the order of the class and summed, then the result is in each case the order of the group. But if the corresponding numbers for two different rows are multiplied, and then multiplied by the order of the class and summed, the result is always zero. Similar results hold for the columns.

Any representation of a group is equivalent to a direct sum of the representations obtained from the group algebra, each one being omitted or repeated any number of times. The manner of this equivalence can be determined by comparing the spurs of the matrices of the representation with the simple characters.

Thus for the symmetric group on three symbols there is a representation for which the elements I, $(\alpha\,\beta\,\gamma)$, $(\alpha\,\gamma\,\beta)$ correspond to the matrix $\begin{bmatrix} 1 & \\ & 1 \end{bmatrix}$, while the elements $(\alpha\,\beta)$, $(\beta\,\gamma)$, $(\alpha\,\gamma)$ correspond to $\begin{bmatrix} & 1 \\ 1 & \end{bmatrix}$. For the three classes, the spurs of the representations are respectively

$$2,\ 0,\ 2,$$

and this *compound character* is clearly equivalent to the sum of the first two characters given in the above table of characters. Hence the representation is equivalent to the direct sum of these two representations. Transforming by the fixed matrix $\begin{bmatrix} 1, & 1 \\ 1, & -1 \end{bmatrix}$, we obtain an equivalent representation of the form $\begin{bmatrix} 1 & \\ & 1 \end{bmatrix}$ for I, $(\alpha\,\beta\,\gamma)$, $(\alpha\,\gamma\,\beta)$, and $\begin{bmatrix} 1 & \\ & -1 \end{bmatrix}$ for $(\alpha\,\beta)$,$(\beta\,\gamma)$, $(\alpha\,\gamma)$. This is clearly the direct sum of the first two representations.

A matrix representation of a group is necessarily also a matrix representation of any sub-group. Hence a character of a group is necessarily a character, either simple or compound, of any sub-group. From the table of characters of a group, then, it is usually possible to determine the table of characters of any sub-group, firstly by taking the given characters as compound characters of the sub-group, and then separating these into simple characters by the use of the orthogonal properties.

Many methods are known for determining the table of characters of a symmetric group. By the above method the characters of the sub-groups also may be found. Many properties of a group may then be determined, almost by inspection, from the table of characters.

Thus the symmetric group on eight symbols is known to have a sub-group of order 1 3 4 4. The table of characters of this sub-group which can be found as above described from the characters of the symmetric group on eight symbols is as follows:

Cycles of Class	1^8	$1^4 2^2$	2^4	2^4	$1^2 3^2$	$1^2 24$	26	17	17	4^2	4^2
Order of Class	1	42	42	7	224	168	224	192	192	168	84
(*a*)	1	1	1	1	1	1	1	1	1	1	1
(*b*)	6	2	2	6	0	0	0	-1	-1	0	2
(*c*)	7	3	-1	-1	1	1	-1	0	0	-1	-1
(*d*)	14	2	2	-2	-1	0	1	0	0	0	-2
(*e*)	21	1	-3	-3	0	-1	0	0	0	1	1
(*f*)	7	-1	3	-1	1	-1	-1	0	0	1	-1
(*g*)	21	-3	1	-3	0	1	0	0	0	-1	1
(*h*)	7	-1	-1	7	1	-1	1	0	0	-1	-1
(*j*)	8	0	0	8	-1	0	-1	1	1	0	0
(*k*)	3	-1	-1	3	0	1	0	$\frac{1}{2}(-1-\sqrt{-7})$	$\frac{1}{2}(-1+\sqrt{-7})$	1	-1
(*l*)	3	-1	-1	3	0	1	0	$\frac{1}{2}(-1+\sqrt{-7})$	$\frac{1}{2}(-1-\sqrt{-7})$	1	-1

Notice that for the characters (*a*), (*b*), (*h*), (*j*), (*k*), (*l*) the characteristics of the fourth class are the same as those of the identity. This is an infallible test for an invariant sub-group. In these representations, those elements from the first and fourth classes correspond to the identity matrix, and hence must form an invariant sub-group of order 8. This group is easily seen to be solvable if, and only if, the quotient group of order $1344/8 = 168$ is solvable. The six characters mentioned give representations of the quotient group, and the table of characters of this quotient group of order 168 are easily determined as follows:

Class	*A*	*B*	*C*	*D*	*E*	*F*
Order	1	21	42	56	24	24
(*a*)	1	1	1	1	1	1
(*b*)	6	2	0	0	−1	−1
(*h*)	7	−1	−1	1	0	0
(*j*)	8	0	0	−1	1	1
(*k*)	3	−1	1	0	$\frac{1}{2}(-1-\sqrt{-7})$	$\frac{1}{2}(-1\sqrt{-7})$
(*l*)	3	−1	1	0	$\frac{1}{2}(-1\sqrt{-7})$	$\frac{1}{2}(-1-\sqrt{-7})$

Apart from the first character (*a*), there is no class which has any characteristic equal to the corresponding characteristic of the identity. Hence there is no self-conjugate sub-group. The group is not solvable, and neither is the group of order 1344.

The symmetric group on eight symbols has another group of order 1152. By examining the table of characters in the same way it is not difficult to show that there is a sub-group of order 288 of which the quotient group is solvable. Treating this group in the same way, there is an invariant sub-group of order 144. This in turn has an invariant sub-group of order 16 which is solvable, and for which the quotient group is solvable. Thus the group of order 1152 is solvable.

This is the largest solvable group of the symmetric group of order 8! Given an eighth degree equation, there is a resolvent equation of degree 35 corresponding to this sub-group. If it should happen that this resolvent has a rational root, then the eighth degree equation would be solvable by the repeated extraction of roots. Otherwise it would not follow that the equation was not solvable, because there exist also other solvable sub-groups which are not sub-groups of the 1152 group. Further examination for the other solvable sub-groups would be required.

CHAPTER XV

THE SYMMETRIC GROUP

THE characters of the symmetric groups have been obtained by Frobenius as coefficients in certain expansions. They have also been obtained explicitly as far as the symmetric group of order 13! Further, the actual matrix representations of the symmetric group have been obtained by Alfred Young. Thus knowledge concerning the symmetric group is fairly extensive.

The applications of these results do not lag behind the knowledge. Frobenius' formula for the symmetric group characters leads directly to the definition of Schur-functions which not only extend our knowledge of symmetric functions, but do themselves form the characters of continuous groups of matrix transformations. In this role they shed light on the structure of tensors, and have important applications to the theory of invariants.

Young's representations of the symmetric group are in every respect complimentary to the theory of Schur-functions and have extensive use concerning tensors, and in invariant theory.

The symmetric group is intimately connected with the theory of symmetric functions, and it is as well to begin with some account of these.

We consider the symmetric functions of a set of n quantities

$$\alpha_1, \alpha_2, \ldots\ldots, \alpha_n.$$

Only those functions are considered which are polynomials in the α's. If the function includes a term $\alpha_1^a\, \alpha_2^b\, \alpha_3^c \ldots$, it will also include all the different terms of the same form that can be obtained by taking different suffixes for the α's. The sum of such different terms is denoted by $\Sigma\, \alpha_1^a\, \alpha_2^b\, \alpha_3^c \ldots$ and is called a *monomial symmetric function*. Clearly, all polynomial symmetric functions can be expressed linearly in terms of these. A particular case of these is given by the *power sums* such as $\Sigma\, \alpha_1^r$ which is denoted by S_r. Any symmetric function can be expressed *algebraically* in terms of the S_r's. Thus to express $\Sigma\, \alpha_1^2\, \alpha_2\, \alpha_3$ in terms of the S_r's,

$$\Sigma\, \alpha_1^2\, \Sigma\, \alpha_1\, \alpha_2 = \Sigma\, \alpha_1^3\, \alpha_2 + \Sigma\, \alpha_1^2\, \alpha_2\, \alpha_3,$$

$$S_1^2 = (\Sigma\, \alpha_1)^2 = \Sigma\, \alpha_1^2 + 2\, \Sigma\, \alpha_1\, \alpha_2 = S_2 + 2\, \Sigma\, \alpha_1\, \alpha_2,$$

$$S_1\, S_3 = \Sigma\, \alpha_1\, \Sigma\, \alpha_1^3 = \Sigma\, \alpha_1^4 + \Sigma\, \alpha_1^3\, \alpha_2 = S_4 + \Sigma\, \alpha_1^3\, \alpha_2.$$

Hence $\Sigma\, \alpha_1^2\, \alpha_2\, \alpha_3 = S_2\, \Sigma\, \alpha_1\, \alpha_2 - \Sigma\, \alpha_1^3\, \alpha_2$

$$= \tfrac{1}{2} S_2 (S_1^2 - S_2) - S_1\, S_3 + S_4.$$

The α's may be considered as the roots of an equation

$$(x - \alpha_1)(x - \alpha_2) \ldots (x - \alpha_n) \equiv x^n - a_1\, x^{n-1} + a_2\, x^{n-2} \ldots \pm a_n = 0.$$

The coefficients are called *elementary symmetric functions*, and are expressed in terms of the α's as follows:

$$a_1 = \Sigma\, \alpha_1,\; a_2 = \Sigma\, \alpha_1\, \alpha_2,\; a_3 = \Sigma\, \alpha_1\, \alpha_2\, \alpha_3,\; \ldots\, ,$$

$$a_n = \alpha_1\, \alpha_2 \ldots \alpha_n.$$

Any symmetric function can be expressed algebraically in terms of the a_r's, just as it can be in terms of the S_r's. The a_r's are algebraically independent, since they can be chosen arbitrarily and the α's will then be the corresponding roots of the equation.

It is more convenient, however, to consider the equation whose roots are the reciprocals of the α's, i.e.

$$f(x) \equiv (1 - \alpha_1 x)(1 - \alpha_2 x) \ldots (1 - \alpha_n x) \equiv 1 - a_1 x + a_2 x^2 - \ldots \pm a_n\, x^n = 0,$$

so that a_r is the coefficient of x^r. Then by formal division $1/f(x)$ can be expressed as an infinite series of ascending powers of x in the form

$$1/f(x) = 1 + h_1\, x + h_2\, x^2 + \ldots + h_r\, x^r \ldots\ldots$$

For sufficiently small x the infinite series does converge to the value of $1/f(x)$, but convergence is not here in question, the formal division can be performed irrespective of convergence. The coefficients h_r constitute another type of symmetric function called the *homogeneous product sums.* In terms of the roots h_r is equal to the sum of all the monomial symmetric functions of degree r in the α's. Thus

$$h_1 = \Sigma\, \alpha_1,\; h_2 = \Sigma\, \alpha_1^2 + \Sigma\, \alpha_1\, \alpha_2,\; h_3 = \Sigma\, \alpha_1^3 + \Sigma\, \alpha_1^2\, \alpha_2 + \Sigma\, \alpha_1\, \alpha_2\, \alpha_3, \text{ etc.}$$

Known formulae express the h_r's in terms of either the a_r's or the S_r's, and conversely either of the latter can be expressed in terms of the others or in terms of the h_r's.

These symmetric functions have close connections with the symmetric groups. Thus if h_r is expressed in terms of the S_r's to give

$$r\,!\,.\,h_r = \Sigma g_\rho \, S_1^a \, S_2^b \, S_3^c \ldots.,$$

then g_ρ is equal to the order of the class

$$\rho = (1^a \, 2^b \, 3^c \ldots)$$

of the symmetric group. This expression corresponds to an idempotent element of the group algebra belonging to the sub-algebra with the character which is unity for each group element.

If $(\lambda) \equiv (\lambda_1, \lambda_2, \ldots, \lambda_i)$ denotes a partition of n, then there is an idempotent element of the group algebra obtained from the sum of the symmetric group elements on the first λ_1 symbols, times the sum of the symmetric group elements on the next set of λ_2 elements, and so on, the product being divided by $\lambda_1!\,\lambda_2!\ldots$. To this idempotent element, which is not primitive but has components in different sub-algebras, there corresponds the symmetric function

$$n\,!\,h_{\lambda_1} h_{\lambda_2} \ldots h_{\lambda_i} = \Sigma g_\rho \, \phi_\rho^{(\lambda)} \, S_1^a \, S_2^b \, S_3^c \ldots,$$

where the coefficients $\phi_\rho^{(\lambda)}$ are compound characters, being the sum of simple characters corresponding to the various sub-algebras in which the above idempotent has components.

By a manipulation of the symmetric functions it can be shown that the same coefficient $\phi_\rho^{(\lambda)}$ is the coefficient of $a_1^{\lambda_1} a_2^{\lambda_2} a_3^{\lambda_3} \ldots$ in the expansion of the product $S_1^a \, S_2^b \, S_3^c \ldots$.

$$S_1^a \, S_2^b \, S_3^c \ldots = \Sigma \, \phi_\rho^{(\lambda)} \, a_1^{\lambda_1} a_2^{\lambda_2} a_3^{\lambda_3} \ldots$$

The coefficients of the various monomial symmetric functions on the right give a series of compound characters, as many as there are partitions of n, from which the simple characters can be deduced by the use of the orthogonal properties of the characters.

However, Frobenius went one step further by showing how the simple characters could be deduced in the general case. His formula involves the product of the differences of the a_i's which is denoted by $\Delta\,(a)$ and which can be expressed as follows:

$$\Delta\,(a) = \prod_{i<j} (a_i - a_j) = \Sigma \pm a_1^{n-1} a_2^{n-2} a_3^{n-3} \ldots a_{n-1}.$$

The summation on the right is taken with respect to all permutations of the suffixes of the a's, the minus sign being taken for a negative permutation, i.e. one formed by an odd number of interchanges.

If the product $S_1^a S_2^b S_3^c \ldots$ is multiplied by $\Delta(a)$, then an alternating function of the a's is obtained which is changed in sign by an interchange of two a's. Hence in the expansion in terms of the a's, in the terms which arise all the a's must appear with different indices. Putting the indices of a typical term in descending order, these can be expressed as

$$\lambda_1+n-1,\ \lambda_2+n-2,\ \lambda_3+n-3,\ \ldots,\ \lambda_{n-1}+1,\ \lambda_n,$$

where

$$\lambda_1 \geqslant \lambda_2 \geqslant \ldots . \geqslant \lambda_n.$$

The coefficient of such a term must be a linear combination of the characters with integral coefficients. By a rather involved procedure using the orthogonal properties of the characters and utilizing doubly infinite series Frobenius demonstrated that each coefficient must be a simple character, and that the characters so obtained were all distinct and formed a complete set of characters.

The characters are associated with the partitions (λ) of n. Frobenius' formula then states that if the characteristic corresponding to the partition (λ) of the class $\rho = (1^a\, 2^b\, 3^c \ldots)$ is $\chi_\rho^{(\lambda)}$, then

$$S_1^a S_2^b S_3^c \ldots \Delta(a) = \Sigma \pm \chi_\rho^{(\lambda)}\, a_1^{\lambda_1+n-1}\, a_2^{\lambda_1+n-2} \ldots$$
$$\ldots a_n^{\lambda_n}.$$

The formula is also true if the number n of the quantities a_i is less than the degree m of the symmetric group, but gives in this case only those characters which correspond to partitions of m into less than or equal to n parts.

The character corresponding to the partition (n) is unity for every element of the group. The character corresponding to the partition (1^n) is 1 for the positive elements, and -1 for the negative elements of the group.

Another basis for the expression of symmetric functions

is given by the Schur-functions or, as they are shortly termed, S-functions. Corresponding to a partition $(\lambda) \equiv (\lambda_1, \lambda_2, \ldots, \lambda_i)$ of n, the S- function $\{\lambda\} \equiv \{\lambda_1, \ldots, \lambda_i\}$ is defined by the formula

$$n!\{\lambda\} = \Sigma h_\rho \chi_\rho^{(\lambda)} S_1^a S_2^b S_3^c \ldots .$$

where ρ denotes the class $(1^a 2^b 3^c \ldots)$. Known formula then show that $\{n\} = h_n$ and $\{1^n\} = a_n$.

Frobenius' formula, taken together with the orthogonal properties of the characters, leads to the expression for the S-function as the quotient of two alternating functions of the α's. Thus

$$\{\lambda\} = (\Sigma \pm \chi_1^{\lambda_1+n-1} \alpha_2^{\lambda_2+n-2} \ldots \alpha_n^{\lambda_n}) \div (\Sigma \pm \alpha_1^{n-1} \alpha_2^{n-2} \ldots \alpha_{n-1}).$$

The S-functions can also be expressed as determinants whose elements are the symmetric functions h_r. Thus

$$\{3\,2\,2\} = \begin{vmatrix} h_3, & h_4, & h_5 \\ h_1, & h_2, & h_3 \\ 1, & h_1, & h_2 \end{vmatrix}$$

Notice that the suffixes of the leading diagonal elements are 3, 2, 2, and that the suffixes increase by unity passing from one column to the next.

The partition (3 2 2) is associated with the following graph of nodes:

• • •
• •
• •

If we read in this graph columns instead of rows, we obtain the conjugate partition (3 3 1). The expression for $\{3\,2\,2\}$ in terms of the a_r's is related to this conjugate partition and is

$$\{3\,2\,2\} = \begin{vmatrix} a_3, & a_4, & a_5 \\ a_2, & a_3, & a_4 \\ 0, & 1, & a_1 \end{vmatrix}$$

Every symmetric function of the α's of degree m in these can be expressed linearly in terms of the S-functions corresponding to partitions of m. If the degree m exceeds the number n of the α_i, then all S-functions corresponding to partitions of m into more than n parts are identically zero. Otherwise the S-functions are linearly independent.

The product of two S-functions corresponding to partitions of p and q respectively, can be expressed as a sum with integral coefficients of S-functions corresponding to partitions of $(p + q)$.

The importance of S-functions, however, lies not in the rather pretty theory of symmetric functions to which they lend themselves, but to their role as characters of the continuous group of matrix transformations. This aspect will be discussed in another chapter.

Young's explicit form for the matrix representations of the symmetric group will be described with reference to the representation corresponding to the partition (3 2) of the symmetric group on five symbols α, β, γ, δ, ϵ.

Take the node graph of the partition and replace the nodes by the five symbols in any order to form a *Young Tableau*, as e.g.

$$\begin{pmatrix} \alpha & \beta & \gamma \\ \delta & \epsilon & \end{pmatrix}$$

Take the sum of the operations of the symmetric group of permutations on each row, to form the product

$$P = [I + (\alpha\,\beta) + (\beta\,\gamma) + (\alpha\,\gamma) + (\alpha\,\beta\,\gamma) + (\alpha\,\gamma\,\beta)]\,[I + (\delta\,\epsilon)].$$

Ignoring for the moment the factors $1/3\,!$ and $1/2\,!$, this is the product of two mutually commuting idempotent elements of the group algebra, and hence is itself a numerical multiple of an idempotent.

The symmetric groups correspond respectively to h_3 and h_2. Hence the product corresponds to

$$h_3\,h_2 = \{5\} + \{41\} + \{32\}.$$

This implies that P can be expressed as the sum of three irreducible idempotents (with numerical coefficients) in the algebras corresponding to the partitions (5), (41), (32).

Now take the sum of the operations of the symmetric group on the symbols in each column, but with a minus sign attached to each negative permutation. The product of these is denoted by

$$N = [I - (\alpha\,\delta)]\;[I - (\beta\,\epsilon)].$$

Then N corresponds to the product

$$a_2^2\, a_1 = \{32\} + \{31^2\} + 2\{2^2 1\} + 2\{21^3\} + \{1^5\}$$

The only S-function in common in the expansions of $h_3\, h_2$ and $a_2^2\, a_1$ is $\{32\}$. Hence the product $P\,N$ will have zero in every other sub-algebra except that corresponding to $\{32\}$. It follows that the product PN is a multiple of a primitive idempotent of this sub-algebra.

Corresponding to the partition (32) there are 5! Young Tableaux according to the possible arrangements of the symbols. These are not independent, but Young has shown that there are exactly the same number of *standard* tableaux as there are rows or columns in the representation. A standard tableau is one in which the order of the symbols in each row or column follows the pre-assigned order. Thus taking the order of the symbols to be α, β, γ, δ, ϵ, the standard tableaux corresponding to (32) are

$$\begin{pmatrix} \alpha & \beta & \gamma \\ \delta & \epsilon & \end{pmatrix}, \begin{pmatrix} \alpha & \beta & \delta \\ \gamma & \epsilon & \end{pmatrix}, \begin{pmatrix} \alpha & \beta & \epsilon \\ \gamma & \delta & \end{pmatrix}, \begin{pmatrix} \alpha & \gamma & \delta \\ \beta & \epsilon & \end{pmatrix}, \begin{pmatrix} \alpha & \gamma & \epsilon \\ \beta & \delta & \end{pmatrix}$$

Denoting the corresponding products from the rows and the columns by P_i, N_i, $i = 1, 2, 3, 4, 5$, the numerical coefficients are each $f^{(\lambda)}/n! = 5/120 = 1/24$. Then $P_i\, N_i/24$ is a primitive idempotent.

If $P_i\, N_i\, P_j\, N_j$ were equal to zero in every case we could put $e_{ii} = P_i\, N_i/24$ and thus obtain the matrix representation corresponding to the sub-algebra. If $j < i$ then $N_i\, P_j = 0$ for there will be two symbols in the same row in the j-th tableau and in the same column in the i-th tableau. Thus N_4 has a factor $I - (\alpha\,\beta)$ while P_2 has a factor $I + (\alpha\,\beta)$. Since $[I - (\alpha\,\beta)]\,[I + (\alpha\,\beta)] = 0$, then $N_4\, P_2 = 0$.

If, however, $i < j$ this may not happen. It fails in fact in the one case $N_1\, P_5$. It is necessary then to introduce a third factor $M_1 = I - P_5\, N_5/24$. Putting $M_2 = M_3 = M_4 = M_5 = 1$, the matrix representation is given by

$$e_{ii} = P_i\, N_i\, M_i/24,$$
$$e_{ij} = P_i\, \sigma_{ij}\, N_j\, M_j/24.$$

Where σ_{ij} is the operation of the group which transforms the j-th tableau into the i-th tableau.

This procedure gives the complete matrix representation of the symmetric group on n symbols corresponding to any partition of n.

Young has used the representation very extensively to analyse the various concomitants obtained in invariant theory.

CHAPTER XVI

CONTINUOUS GROUPS

THE complex number $x + i\,y$ is represented on a diagram called the Argand Diagram by the point P whose rectangular Cartesian co-ordinates are (x, y). The length $OP = \sqrt{(x^2 + y^2)}$ is called the modulus of $(x + i\,y)$. If $OP = 1$ and the angle XOP is θ, the complex number is denoted by $e^{i\theta}$. Multipli-

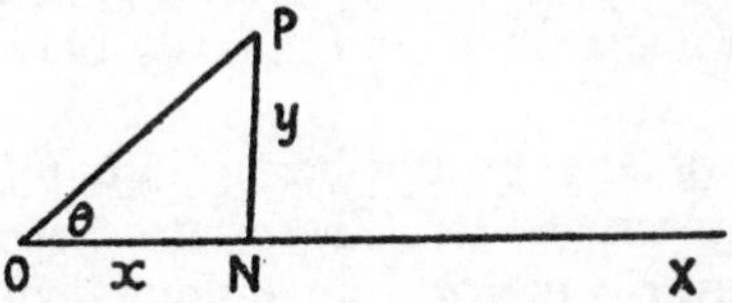

cation by $e^{i\theta}$ corresponds to a rotation through an angle θ in a counter-clockwise direction. Consequently multiplication n times by θ corresponds to a rotation through an angle $n\,\theta$, and

$$(e^{i\theta})^n = e^{in\theta}.$$

Now consider a regular hexagon $ABCDEF$. The hexagon can be rotated through an angle of either 60°, 120°, 180°, 240°, 300°, so as to leave it in coincidence with its original

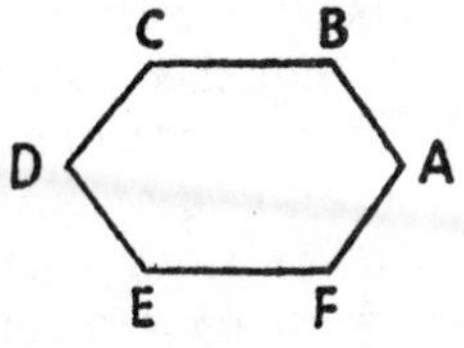

shape. These rotations form a group, a cyclic group of order 6 called the group of rotations of a regular hexagon. Each of the six rotations of the group forms a class by itself. Hence there are six classes and six representations.

There is a representation in which a rotation through an angle θ corresponds to the complex number $e^{i\theta}$. Another representation makes the rotation correspond to $e^{2i\theta}$, a third $e^{3i\theta}$, and then for the fourth and fifth $e^{4i\theta}$ and $e^{5i\theta}$. Since $e^{6i\theta} = 1$, the sixth representation in this sequence makes each rotation correspond to unity, and gives the unitary representation common to every group.

Now if the regular hexagon is replaced by a regular n-sided figure, there will be similar representations in which a rotation through an angle θ is represented by respectively $e^{i\theta}$, $e^{2i\theta}$, $e^{3i\theta}$, ... , $e^{(n-1)i\theta}$, $e^{ni\theta} = 1$.

Proceeding to the limit as n tends to infinity we obtain the group of rotations of a circle. The elements of this group, which are the same as the classes, form a continuous manifold, since a rotation could vary continuously between zero and an angle θ.

The order of the group being infinite, there exist an infinity of representations. These are not connected continuously, however, but form an infinite sequence of distinct representations, a rotation through an angle θ being represented respectively by

$$1, e^{i\theta}, e^{2i\theta}, e^{3i\theta}, \ldots\ldots, e^{ni\theta}, \ldots\ldots$$

Such a group is called a *continuous group*. The theory of matrix representations of a finite group extends quite naturally to the infinite continuous groups. The example discussed, of rotations in a plane, is a commutative group and all the representations are of degree one, but the group of rotations in three or more dimensions may be considered similarly, as well as the group of all non-singular linear transformations on n variables, and an infinite sequence of true matrix representations will be obtained. In the case of the rotation groups, as well as extensions of these to complex transformations called unitary groups, the orthogonal properties of the characters still hold, if the summation over elements of the group is replaced by an integration, which is a continuous form of summation, over the group manifold.

The simplest continuous group is the group of non-singular linear transformations on a set of n variables. The element of the group is the matrix of the transformation.

Since two matrices with the same characteristic equation can be transformed into one another, the class of conjugate elements is specified by the characteristic equation of the matrix. Suppose that this equation is

$$\lambda^n - a_1 \lambda^{n-1} + a_2 \lambda^{n-2} \ldots \pm a_n = 0.$$

The coefficients in the characteristic equation are regarded as symmetric functions of the latent roots of the matrix.

The matrices themselves form one matrix representation of the group, and the corresponding character is the spur of this matrix, which is $a_1 = \{1\}$. Other matrix representations are obtained as follows.

Suppose that the n variables $x_1, x_2, \ldots, x_n$ are transformed by the given matrix. Then the $\frac{1}{2}n(n+1)$ powers and products of degree two, $x_1^2, x_2^2, \ldots, x_n^2, x_1 x_2, x_1 x_3, x_2 x_3, \ldots\ldots$ will be transformed by a matrix of degree $\frac{1}{2}n(n+1)$, which is called the second induced matrix. The spur of this matrix is the symmetric function $h_2 = \{2\}$, which therefore forms a character of the group. Similarly for the powers and products of degree 3, 4, 5, etc., the matrix of transformation is the third, fourth, fifth, etc., induced matrices, and the corresponding characters are $\{3\}$, $\{4\}$, $\{5\}$, etc.

Again, suppose that the variables $y_1, y_2, \ldots, y_n$ are transformed by the same matrix of transformation as $x_1, x_2, \ldots, x_n$. Then the n^2 products $x_i y_j$ will be transformed by a matrix with n^2 rows and columns which is equivalent to the *direct product* of the original matrix with itself. Its spur will be a_1^2.

The n^2 linearly independent terms $x_i y_j$ can be separated into those which are symmetric with respect to the n's and the y's, and those which are anti-symmetric. Thus there are the symmetric terms

$$x_i y_i, \; x_i y_j + x_j y_i,$$

which will be transformed by the second induced matrix, and the anti-symmetric terms

$$x_i y_j - x_j y_i,$$

which are transformed by a matrix of $\frac{1}{2}n(n-1)$ rows and columns which is called the *second compound* matrix. The spur of this matrix is a_2.

Hence the direct product of the original matrix with itself is equivalent to the direct sum of the second compound and the second induced matrices. This corresponds to the equation connecting the spurs, which are the characters of the representations

$$a_1^2 = h_2 + a_2,$$

or $$\{1\}^2 = \{2\} + \{1^2\}.$$

Of degree three, there are the matrices of transformation on the terms

$$x_i^3,\ x_i^2\, x_j,\ x_i\, x_j\, x_k,$$

namely the third induced matrix which has the spur $\{3\}$, upon the $\frac{1}{6}\, n(n-1)\,(n-2)$ terms which are the values of the determinants

$$\begin{vmatrix} x_i, & x_j, & x_k \\ y_i, & y_j, & y_k \\ z_i, & z_j, & z_k \end{vmatrix},$$

which is the third compound matrix, having spur $\{1^3\}$, and also some extra terms which are partly symmetric and partly anti-symmetric.

The matrix of transformation on such terms as

$$x_i\,(x_i\, y_j - x_j\, y_i),\ x_i\,(x_j\, y_k - x_k\, y_j)$$

gives a representation of the group whose spur is the S-function $\{21\}$.

If x_i, y_i, z_i are three sets of variables each transformed by the original matrix, the n^3 products $x_i\, y_j\, z_k$ are transformed by a matrix which is the direct product of three matrices, each identical with the original matrix. These terms separate into those symmetric, those anti-symmetric, and the mixed terms of type $\{21\}$, according to the formula

$$\{1\}^3 = \{3\} + \{1^3\} + 2\,\{21\}.$$

The numerical coefficient 2 implies that there are two distinct sets of terms which are transformed by the same matrix of transformation of type $\{21\}$.

Similarly for degree m there is a matrix of transformation corresponding to each partition of m into not more than n parts.

The anti-symmetric terms of degree n in n variables have but a single term which is multiplied in a transformation by the determinant of the transformation. Thus the S-function $\{1^n\}$ is the character of a unary representation of the group of transformations in which each transformation is represented by its determinant D. Powers of D such as D^r also form a representation, the corresponding character being $\{r^n\}$.

The rules for the multiplication of S-functions show the manner in which the representations combine with one another in forming direct products.

Another group of transformations is the orthogonal group. In rectangular Cartesian co-ordinates the distance d between a point (x, y, z) and the origin is given by

$$d^2 = x^2 + y^2 + z^2.$$

If this distance is to be independent of the particular system of co-ordinates used, only those transformations may be used which leave the quadratic form $x^2 + y^2 + z^2$ invariant.

In the general case any quadratic form may be left invariant. This quadratic form is called the *metric*. In terms of tensors, for a metric in the n variables $x^1, x^2, \ldots, x^n$, the form is usually taken as $g_{ij}\, x^i\, x^j$. Transformations which leave this form invariant are said to form the general orthogonal group.

This includes space time transformations of the Lorentz group for which we may take

$$g_{00} = -1, g_{11} = g_{22} = g_{33} = 1, g_{ij} = 0\ (i \neq j),$$

so that the metric is

$$(x^1)^2 + (x^2)^2 + (x^3)^2 - (x^0)^2$$

where $x^0 = c\, t$, t being the time and c the velocity of light.

The tensor g_{ij} is a symmetric tensor of type $\{2\}$. There is a contragredient tensor g^{ij} such that

$$g^{ij}\, g_{ij} = n$$

and

$$g^{ij}\, g_{ik} = 1 \text{ if } j = k,$$
$$= 0 \text{ if } j \neq k.$$

These tensors can be used for raising or lowering the suffixes. Thus if a_i is a cogredient tensor of rank 1,

$$b^i = g^{ij}\, a_j$$

is the corresponding contragredient tensor. The matrix of transformation is clearly the same for each.

Now if a_{ij} is a tensor of type $\{2\}$, that is, a symmetric tensor of rank 2, so that $a_{ij}\, x^i\, x^j$ is a quadratic form, then this quadratic form will possess an invariant

$$g^{ij}\, a_{ij}.$$

Hence for the orthogonal group the second induced matrix

of the matrix of transformation is reducible. The equation for the corresponding characters is

$$\{2\} = [2] + [0].$$

The symbol $[0] = 1$ is the character of the representation which is unity for each element. The symbol [2] denotes the character of the orthogonal group corresponding to a quadratic from which the invariant term has been removed.

The representations of the full linear group, which are called *invariant matrices*, give representations also of the orthogonal group, but in general, like the quadratic, they are reducible.

For $n = 2\nu$ or $n = 2\nu + 1$ variables, the partitions corresponding to less than or equal to ν parts separate into representations corresponding to smaller partitions together with one new representation of which the character is denoted by the partition placed inside square brackets. Hence there is one representation of the orthogonal group corresponding to each partition into not more than ν parts.

Various formulae enable one to express the *S*-functions in term of the orthogonal group characters.

Thus an important tensor in connection with the theory of Relativity is called the Riemann Chrystofel Tensor. It is of type $\{2^2\}$ and contains 20 terms, 20 being the degree of the invariant matrix corresponding to $\{2^2\}$ of a four-rowed matrix. Expressing $\{2^2\}$ as a sum of orthogonal group characters, the equation is

$$\{2^2\} = [2^2] + [2] + [0].$$

Correspondingly the Riemann Chrystofel tensor separates into three parts, which include the conformal curvature tensor of type $[2^2]$, the Ricci tensor G_{ij} of type [2] and the curvature G which is a scalar and corresponds to [0].

Invariant matrices corresponding to partitions into more than ν parts, of an orthogonal matrix, separate into representations corresponding to partitions of smaller numbers. For example in 4 variables

$$\{211\} = [2] + [1^2].$$

Besides the representations described, which are called *true* representations, there exist also for the orthogonal group some two-valued representations called ***spin representations*** or ***spinors***. They are two-valued since there is always an ambiguity of sign concerning them.

For $2\nu+1$ variables it requires a set of $(2\nu+1)$ matrices $E_1, E_2, \ldots, E_{2\nu+1}$ which satisfy

$$E_i^2 = \pm 1,\ E_i E_j + E_j E_i = 0.$$

The minimum number of rows necessary for this to be possible is 2^ν, and the E_i's are taken to be 2^ν-rowed matrices.

Then, for any orthogonal transformations

$$X_i' = a_{ij} X_j$$

it is possible to find a matrix T such that

$$T^{-1} E_i T = \Sigma\, a_{ij} E_j.$$

The matrices T form a representation of the group. But since

$$(-T)^{-1} E\, (-T) = T^{-1} E_i T,$$

the same orthogonal transformation occurs when T is replaced by $-T$. This is the ambiguity of sign which is always associated with spinors.

The matrices T are said to form the ***basic spin representation*** of the group. It is associated with a character denoted by $[(\frac{1}{2})^\nu]$. Combinations of the basic spin representation and true representations lead to other spin representations whose characters are denoted by partitions in which each part is half an odd integer.

The spin representations for $n = 2\nu$ variables are the same as for $n = 2\nu + 1$. Thus in four dimensions there is a basic spinor of type $[\frac{1}{2}, \frac{1}{2}]$ with four components. Multiplication by a vector of type $[1]$ leads to a reducible spinor in accordance with the equation

$$[\tfrac{1}{2}, \tfrac{1}{2}]\,[1] = [\tfrac{3}{2}, \tfrac{1}{2}] + [\tfrac{1}{2}, \tfrac{1}{2}],$$

the spinor of type $[\frac{3}{2}, \frac{1}{2}]$ having twelve components.

It is fairly easy to perceive the geometrical significance of the true representations. A tensor of type $[1]$ is a vector, or a quantity such as displacement, velocity or force, which has direction as well as magnitude. The square or cube of a vector would form a tensor of type $[2]$ or $[3]$. An element of area would transform like a tensor of type $[1^2]$, an element of volume would be of type $[1^3]$. In three dimensions an element of area is perpendicular to a given direction and hence may be regarded as a vector in this perpendicular direction, provided that the group is orthogonal, and thus $[1^2] = [1]$. Under the full linear group perpendicularity is not an invariant property, and $\{1^2\}$ and $\{1\}$ are distinct representations.

But the geometrical significance of the spinors is not at all apparent, and they appear to have a somewhat mysterious nature. They have very important applications in Quantum Theory.

Whereas the orthogonal group is a group which leaves a quadratic form invariant, there is a different group which leaves a linear complex, which is a form of type $\{1^2\}$, invariant. The group is called the *symplectic group*. The group is in many respects similar to the orthogonal group, and the characters have been analysed in a similar manner. It is in many respects simpler than the orthogonal group. Thus there is no analogue of the spinors.

CHAPTER XVII

APPLICATION TO INVARIANTS

IT has been shown how tensors can be used in invariant theory for the construction of invariants and other concomitants. This use, however, leads to many zero expressions and many repetitions of the same concomitant.

The use of Young's representation of the symmetric group can be made the basis of a substitutional analysis which analyses these tensor contractions with exactitude. Young used the method, not in connection with tensors, but relating to certain symbols in a *symbolic method* which have had a widespread use in invariant theory. It is essentially equivalent to the use of tensors, and the same calculation could be expressed in the one form or the other. It will be discussed here in reference to tensors.

Calculation with S-functions gives the number and type of the various concomitants with precision, though the method can prove laborious for high degrees. A suitable combination of the use of S-functions and Young's substitutional methods gives the complete set of linearly independent concomitants of any degree.

Firstly we consider the manner of decomposition of a cogredient tensor of rank r. For simplicity we will consider the case $r = 3$, which illustrates the general case sufficiently.

Since the character of the matrix of transformation of a tensor of rank one is $\{1\}$, the character of the matrix of transformation of a complete tensor of rank r will be $\{1\}^r$. This can be expressed in terms of S-functions by a formula based on the definition of S-functions

$$S_1^a S_2^b S_3^c \ldots = \Sigma \chi_\rho^{(\lambda)} \{\lambda\},$$

so that

$$S_1^r = \{1\}^r = \Sigma f^{(\lambda)} \{\lambda\}$$

where $f^{(\lambda)} = \chi_0^{(\lambda)}$ is the degree of the corresponding character.

For a tensor of rank three this gives

$$\{1\}^3 = \{3\} + 2\{21\} + \{1^3\},$$

and this gives the manner of the separation of a tensor of rank three. To find the explicit form for this separation, Young's substitutional operators must be used.

If the tensor A_{ijk} becomes after a transformation A'_{ijk}, then clearly A_{jik} will become A'_{jik}. Hence any operation of permuting the suffixes of the tensor will be carried over unchanged into the transformed tensor. Similarly any linear combination of the forms obtained by permuting the suffixes will be carried over into the same linear combination of the transformed tensors.

Thus for tensors of rank two, $A_{ij} + A_{ji}$ will become $A'_{ij} + A'_{ji}$, and $A_{ij} - A_{ji}$ will become $A'_{ij} - A'_{ji}$. Thus the symmetric tensor remains symmetric after transformation and the anti-symmetric tensor remains anti-symmetric.

For the tensor of rank r, consider the operations of symmetric group of permutations acting upon the r suffixes. The linear combinations of these operations form the group algebra. The modulus of the group algebra can be expressed as a sum of irreducible idempotents corresponding to the different sub-algebras. Operating on the tensor with the substitutional expressions corresponding to the irreducible idempotents has the effect of separating the tensor into its irreducible components. Corresponding to the sub-algebra with partition (λ), there are $f^{(\lambda)}$ irreducible idempotents, and $f^{(\lambda)}$ separate sub-tensors transform according to the corresponding matrix of transformation.

Thus for $r = 3$ the irreducible idempotents are, for the group of permutations on the suffixes i, j, k, corresponding to the partition (3),

$$\tfrac{1}{6}[I + (ij) + (jk) + (ik) + (ijk) + (ikj)],$$

corresponding to the partition (1^3)

$$\tfrac{1}{6}[I - (ij) - (jk) - (ik) + (ijk) + (ikj)]$$

and corresponding to the partition (21), the pair of idempotents

$$\tfrac{1}{3}[I + (ij) - (ik) - (ijk)]$$
$$\tfrac{1}{3}[I + (ik) - (ij) - (ikj)].$$

Ignoring the numerical factors and operating on the

general tensor of rank three A_{ijk} we obtain the symmetric tensor

$$A_{ijk}+A_{jik}+A_{ikj}+A_{kji}+A_{jki}+A_{kij},$$

the alternating tensor

$$A_{ijk}-A_{jik}-A_{ikj}-A_{kji}+A_{jki}+A_{kij}$$

together with the two tensors of type $\{21\}$

$$A_{ijk}+A_{jik}-A_{kji}-A_{kij},$$
$$A_{ijk}+A_{kji}-A_{jik}-A_{jki}.$$

The application of S-functions to invariant theory is as follows. Suppose first that there are given two ground forms which are respectively a cubic and a quartic in a set of variables. Suppose that they are

$$a_{ijk}\,x^i\,x^j\,x^k,\quad b_{ijkp}\,x^i\,x^j\,x^k\,x^p.$$

Now the matrix of transformation of the coefficients a_{ijk} is the third induced matrix of the matrix of transformation corresponding to the character $\{3\}$. Similarly the coefficients b_{ijkp} correspond to the character $\{4\}$. Hence the products $a_{ijk}\,b_{pqrs}$ form a tensor corresponding to

$$\{4\}\{3\}=\{7\}+\{61\}+\{52\}+\{43\}.$$

Now consider those concomitants of the two ground forms which are linear in the coefficients of each. The coefficients are linear combinations of the products $a_{ijk}\,b_{pqrs}$. Hence the coefficients in each concomitant must form one of the tensors of type $\{7\}$, $\{61\}$, $\{52\}$, $\{43\}$. It follows that there are exactly four concomitants linear in each of these two ground forms, and these are forms of type $\{7\}$, $\{61\}$, $\{52\}$ and $\{43\}$ respectively.

The actual concomitants can be constructed as follows. The rules for multiplying S-functions depend on the construction of tableaux corresponding to the S-functions in the product using the symbols from tableaux corresponding to the S-functions to be multiplied. For the product $\{4\}\{3\}$, from the tableaux

$$(\alpha_1\,\alpha_2\,\alpha_3\,\alpha_4),\ (\beta_1\,\beta_2\,\beta_3),$$

the following tableaux are built:

$$(\alpha_1\,\alpha_2\,\alpha_3\,\alpha_4\ \beta_1\,\beta_2\,\beta_3),\quad \begin{pmatrix}\alpha_1\,\alpha_2\,\alpha_3\,\alpha_4\ \beta_1\,\beta_2\\ \beta_3\end{pmatrix},$$

$$\begin{pmatrix}\alpha_1\,\alpha_2\,\alpha_3\,\alpha_4\ \beta_1\\ \beta_2\,\beta_3\end{pmatrix},\quad \begin{pmatrix}\alpha_1\,\alpha_2\,\alpha_3\,\alpha_4\\ \beta_1\,\beta_2\,\beta_3\end{pmatrix}.$$

From these tableaux the concomitants can be written down directly. The variables are taken as x^i, y^i which are two contragredient tensors of rank one. For a tensor of type $\{1^2\}$ we obtain the anti-symmetric tensor

$$x^{ij} = x^i y^j - x^j y^i.$$

Then for any tableau in the product, when an α and a β appear in the same column, a contraction is effected between a suffix of a_{ijk}, a suffix of b_{pqrs}, and a tensor of the form x^{ij}. When a symbol appears in a column by itself the corresponding suffix is contracted with x^i. Thus the concomitants are

$$a_{ijk}\, b_{pqrs}\, x^i\, x^j\, x^k\, x^p\, x^q\, x^r\, x^s,\quad a_{ijk}\, b_{pqrs}\, x^{ip}\, x^j\, x^k\, x^q\, x^r\, x^s,$$
$$a_{ijk}\, b_{pqrs}\, x^{ip}\, x^{jq}\, x^k\, x^r\, x^s,\quad a_{ijk}\, a_{pqrs}\, x^{ip}\, x^{jq}\, x^{kr}\, x^s.$$

The concomitants that are linear in any number of ground forms of any types are found similarly.

For the concomitants of degree two or more in a single ground form the procedure is not so simple. It involves a third type of combination of *S*-functions denoted by $\otimes$, beyond the ordinary addition and multiplication.

The invariant matrix of a matrix A corresponding to the partition (λ) is denoted by $A\,\{\lambda\}$. These invariant matrices form a sub-group of the full group of matrices of the same order. Hence the invariant matrix corresponding to the partition (μ) of the invariant matrix $A\,\{\lambda\}$, which is written $[A\,\{\lambda\}]\,\{\mu\}$ will form a representation of the group of matrices A, possibly reducible, and hence will be equivalent to the direct sum of invariant matrices of A corresponding to various partition (ν).

Then we write

$$\{\lambda\} \otimes \{\mu\} = \Sigma\,\{\nu\}.$$

Techniques for the evaluation of such operations have been examined, and the behaviours with respect to sums and products.

Now suppose that there is given a ground form of type $\{\lambda\}$. The coefficients are transformed by the invariant matrix $A\,\{\lambda\}$. The powers and products of the coefficients of degree r are then transformed by the r-th induced matrix of $A\,\{\lambda\}$, or by $[A\,\{\lambda\}]\,\{r\}$. To the various invariant matrices which appear in this induced matrix there correspond concomitants of degree r in the ground form.

Hence if

$$\{\lambda\} \otimes \{r\} = \Sigma \{\nu\},$$

then for each term $\{\nu\}$ in the sum on the right there is a concomitant of type $\{\nu\}$ which is of degree r in the ground form of type $\{\lambda\}$.

For the concomitants of two forms of types $\{\lambda\}$ and $\{\mu\}$ which are of the respective degrees r and s, the appropriate equation is

$$[\{\lambda\} \otimes \{r\}]\, [\{\mu\} \otimes \{s\}] = \Sigma \{\nu\}.$$

The actual concomitants can be found as before by constructing the tableaux which appear in the various products.

Thus, for example, for the cubic in two variables the equations are

$$\{3\} \otimes \{2\} = \{6\} + \{4, 2\},$$
$$\{3\} \otimes \{3\} = \{9\} + \{7, 2\} + \{6, 3\},$$
$$\{3\} \otimes \{4\} = \{12\} + \{10, 2\} + \{9, 3\} + \{8, 4\} + \{6, 6\}.$$

These give the types of concomitant up to degree 4 in the gound-form coefficients.

Notice that there is a concomitant of degree 2 and type $\{6\}$. This is just the square of the ground form and is said to be reducible. Similarly, of degree 3, the term $\{9\}$ is the cube of the ground form, while $\{7, 2\}$ is the product of the ground form and the concomitant of type $\{4, 2\}$.

The only irreducible concomitants are

Degree 2; $\{4, 2\}$. Degree 3; $\{6\ 3\}$. Degree 4; $\{6\ 6\}$.

The corresponding tableaux are

$$\begin{pmatrix} \alpha\,\alpha\,\alpha\,\beta \\ \beta\,\beta \end{pmatrix}, \quad \begin{pmatrix} \alpha\,\alpha\,\alpha\,\beta\,\gamma\,\gamma \\ \beta\,\beta\,\gamma \end{pmatrix}, \quad \begin{pmatrix} \alpha\,\alpha\,\alpha\,\beta\,\gamma\,\gamma \\ \beta\,\beta\,\gamma\,\delta\,\delta\,\delta \end{pmatrix}.$$

Two symbols in the same column can be contracted with the alternating tensor since there are only two variables. Hence the respective forms are a quadratic, a cubic, and an invariant. The actual forms can be written down from the tableaux.

The same manner of procedure may be employed to find the concomitants of a form or a system of forms under the orthogonal group or under the symplectic group. The characters of the respective groups are employed and expressed in terms of S-functions when the above procedure may be adopted.

In mathematical physics, in Quantum Theory and especially in the theory of the atomic nucleus, invariantive methods are extremely powerful. The matrix representations of the full linear and of the orthogonal groups have direct applications. These applications are, however, much too intricate to be described here.

The remarkable feature concerning a considerable proportion of the applications of Algebra is their unexpected and unpredictable nature. The theory of non-commutative algebras and the theory of group representation were studied by pure mathematicians in a disinterested search for pure knowledge. No stretch of the imagination would have conceived, at the times of their discovery, that the former would prove to be the very tool which the physicist required for the elucidation of problems in quantum theory, while the latter would prove to be one of the keys to the elucidation of the atomic nucleus.

For the algebrist to read in other branches of science and of mathematics, and to encounter these applications, it is as if a man travelling far away in a foreign land suddenly and without warning meets a friend from his own country, from his own home town.

INDEX

ALGEBRAS, 101
algebraic field, 41

CEVA's theorem, 83
character, characteristic, 110
compound matrices, 125
concomitants, 79, 95
congruences, 30
continued fractions, 27
continuous groups, 123
co-ordinates:
  Cartesian, 77
  homogeneous, 83
covariant, 95
cross ratio, 84
cycles, 58

DESCARTES, 77
determinant, 89
direct product, 125
division algebra, 103

EQUATIONS:
  cubic, 65
  quartic, 69
  solvable, 67
Euclid's algorithm:
  integers, 23
  polynomials, 41

FACTORIZATION (unique), 25, 41, 52
Fitzgerald contraction, 96
Frobenius:
  algebra, 107
  formula, 118

GALOIS:
  resolvent, 72
  theory, 65
Grassmann algebra, 103
group, 58
  alternating group, 69
  symmetric group, 61
group algebra, 107

IDEAL, 53
induced matrix, 125
invariant, 79, 94
invariant matrices, 125

LATENT roots, 90
Lorentz transformation, 96

MATRIX, 96
  orthogonal matrix, 91
  singular matrix, 88
  symmetric matrix, 91
matrix representations, 109

NILPOTENT, 104
numbers, 17
  complex numbers, 44

ORTHOGONAL:
  group, 127
  matrix, 91

P-ADIC numbers, 55
partial fractions, 25
partition, 62

permutations, 61
polar, 79
pole (of matrix), 90
polynomials, 40
projection, 85

QUADRATIC reciprocity, 34, 51
quaternions, 101

RELATIVITY:
  general, 10
  special, 12, 95
residue, 30
rule and compass constructions, 74

SCHUR functions, S-functions, 119

spinors, 129
spur (of matrix), 110
subgroups, 67
symbolic method, 131
symmetric:
  functions, 115
  group, 61
  matrix, 91
symplectic group, 130

TENSOR, 97
trace, 107
transform, 61
transformation, 86

YOUNG tableau, 120

A CATALOG OF SELECTED
DOVER BOOKS
IN ALL FIELDS OF INTEREST